IMAREST
publications

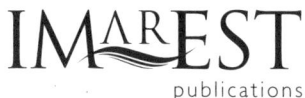

Merchant Ship Naval Architecture

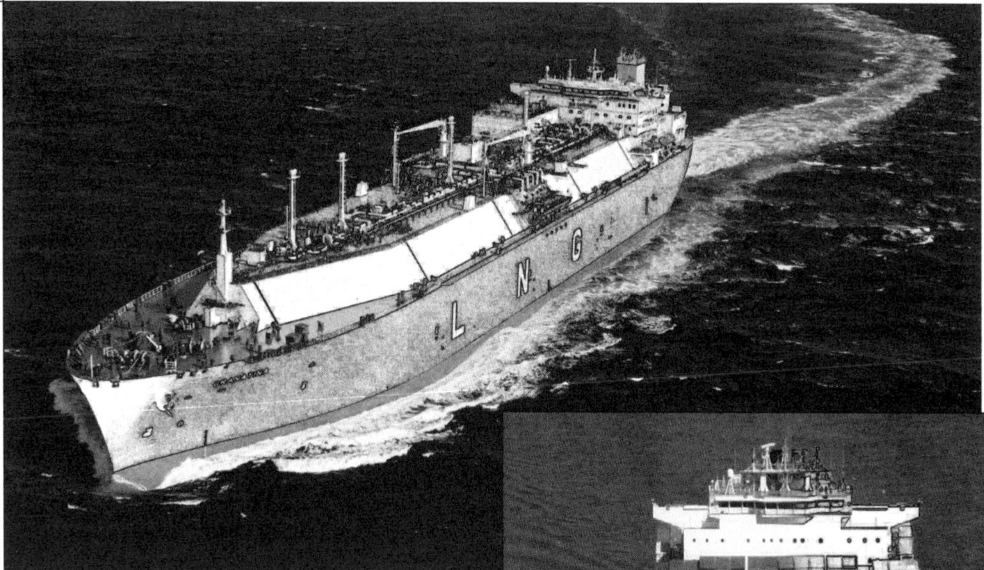

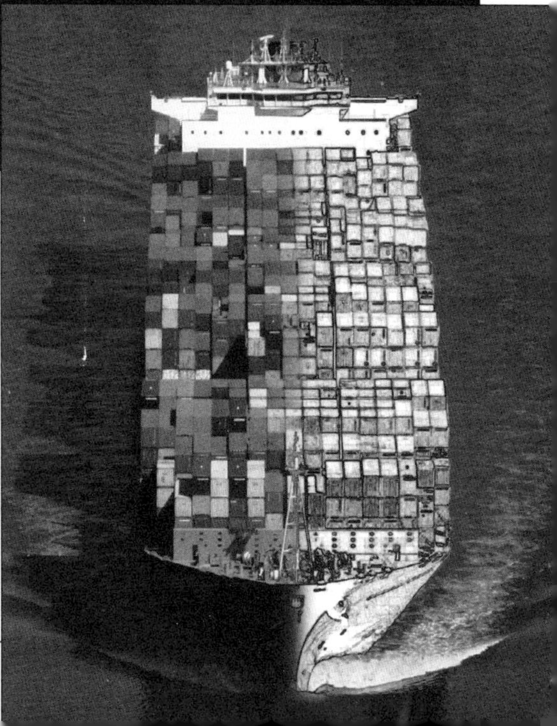

Published by IMarEST
The Institute of Marine Engineering, Science and Technology
80 Coleman Street • London • EC2R 5BJ

A charity registered in England and Wales
Registered Number 212992

Email: publications@imarest.org

www.imarest.org

A copy of the latest IMarEST Publications Catalogue is available to download from
www.imarest.org/publications/
An e-shop facility to purchase books is also available

All rights reserved. No part of this publication may be reproduced or stored in a retrieval system, or transmitted in any form or by any means, electronic, mechanical, photocopying or otherwise, without the prior permission of the copyright holder

A CIP catalogue record for this book is available from the British Library

Copyright © 2006 IMarEST. All rights reserved.
The Institute of Marine Engineering, Science and Technology

ISBN 1-902536-56-8

Preface

This book aims to provide a clear understanding of the principles of naval architecture, giving the reader a basic grounding in the theoretical aspects of the different branches of the subject: resistance and propulsion, structural strength, ship motion and control, stability and trim etc.

The book begins with a brief introduction to modern ship types outlining their principal design features and concludes with an overview of the ship design process in the final chapter. Brief reference to international regulations and the operating environment, the sea, have also been made.

While aimed at undergraduate students pursuing naval architecture, marine engineering or maritime technology courses, this book should also prove useful to seagoing marine engineers and deck officers, when preparing for their certificate of competency examinations.

Computers have made a great impact on all aspects of ship design, construction and operation. Design calculations by computer programs are now a matter of course. Having grasped the fundamentals of naval architecture, readers may wish to experiment with computer programs that are accessible through the internet, gaining a deeper understanding and practice in the subject.

Dr David A Taylor
Dr Alan ST Tang

Acknowledgements

The authors would like to acknowledge the assistance provided by the following companies and organisations:

ABB Azipod Oy
American Bureau of Shipping
Brown Bros & Co Ltd
Lloyd's Register of Shipping
The Royal Institution of Naval Architects
International Maritime Organization
Wolfson Unit MTIA
Institute of Marine Engineering, Science & Technology (IMarEST)

Material has been incorporated in this book, by agreement, that was initially used in *Muckle's Naval Architecture*, 2/e by W Muckle, revised by DA Taylor (Butterworths). Some material and a number of examples have been used, with permission, from *Ships and Naval Architecture* by R Munro-Smith (published by IMarEST).

We would like to thank everyone who has helped us in the writing of this book.

Dr David A Taylor
Dr Alan ST Tang

About the Authors

David A Taylor was a consulting Naval Architect and Marine Engineer in Hong Kong with Habour Craft Services Ltd and the author of the best selling book *Merchant Ship Construction* published by the IMarEST. He served for a period at sea before attending the University of Newcastle upon Tyne, UK, from which he obtained a BSc(Hons) in Marine Engineering and an MSc in Shipbuilding. He worked for some years for Swan Hunter Shipbuilders and lectured at South Shields Marine and Technical College and Hong Kong Polytechnic (now University). He was awarded a PhD by the Department of Maritime Studies and International Transport of the University of Wales Cardiff. He was a Fellow of the IMarEST and a number of engineering institutions in the UK and Hong Kong. David Taylor passed away in December 2003, leaving a legacy of marine textbooks and publications, of which this book published posthumously is his last major work for the students of maritime transport.

Alan ST Tang is a senior lecturer at the Vocational Training Council (VTC), the major vocational education and training provider in Hong Kong. Apart from lecturing, he has led projects in marine related consultancy work and research projects. He has worked on a research project in containership emissions and as a professional writer for the qualification framework in Hong Kong, providing expertise in ship related work. He has published over 30 journal and international conference papers, with topics ranging from high speed craft development, ship design and operation to engineering education and training. He obtained a BSc and PhD from the University of Southampton and an MBA from the University of Birmingham, UK. Prior to joining the VTC, he conducted research in warship stabilisation in the UK. He is a member of RINA.

Contents

Preface i
Acknowledgements ii
About the Authors iii

1 Ship types 1
    1.1 General cargo ships 2
    1.2 Refrigerated cargo ships 3
    1.3 Containerships 3
    1.4 Roll-on roll-off ships 4
    1.5 Barge carriers 5
    1.6 Oil tankers 6
    1.7 Bulk carriers 7
    1.8 Liquefied gas carriers 9
    1.9 Liquefied natural gas carriers 10
    1.10 Liquefied petroleum gas carriers 10
    1.11 Chemical tankers 12
    1.12 Passenger ships 13
    1.13 Fast ferry designs 14
    1.14 Summary 16

2 Defining a ship 17
    2.1 Principal dimensions 17
    2.2 Important features 18
    2.3 Hull form 18
    2.4 Displacement 21
    2.5 Tonnage 21

3 Rules, regulations and legislation 23
    3.1 Classification societies 23
    3.2 National authorities 24
    3.3 IMO 25
    3.4 Safety 26
    3.5 Prevention of pollution 27
    3.6 Fire safety in ships 27
    3.7 The load line rules – freeboard 28
    3.8 Tonnage 29
    3.9 The International Convention on Tonnage Measurement of Ships 29
    3.10 Other tonnage systems 30

4 Calculation of areas and volumes 31
    4.1 Trapezoidal rule 31
    4.2 Simpson's rule 32
    4.3 Application to volumes 34
    4.4 Application to First and Second Moments of Area 34
    4.5 Summary 36

5 Buoyancy, stability and trim ... 39
- 5.1 Buoyancy and displacement ... 39
- 5.2 Bonjean curves ... 40
- 5.3 Transverse stability ... 40
- 5.4 Determining the transverse metacentre ... 41
- 5.5 Metacentric diagram ... 43
- 5.6 Determining the centre of gravity ... 43
- 5.7 Conduct of the inclining experiment ... 44
- 5.8 Operations affecting stability ... 46
- 5.9 Free surface correction ... 46
- 5.10 Use of tank divisions ... 47
- 5.11 Effect of suspended weights on stability ... 48
- 5.12 Transverse movement of weight ... 48
- 5.13 Large angle stability ... 49
- 5.14 Wall-sided ships ... 50
- 5.15 Cross curves of stability ... 53
- 5.16 Curves of statical stability ... 53
- 5.17 Dynamical stability ... 56
- 5.18 Trim ... 56
- 5.19 Effect of adding a small weight ... 57
- 5.20 Hydrostatic curves ... 59
- 5.21 Effect of adding a large weight ... 59
- 5.22 Displacement determination from measured draughts ... 61
- 5.23 Effect of water density changes on draught ... 62
- 5.24 Docking and stability ... 62
- 5.25 Squat ... 64
- 5.26 Flooding and subdivision ... 66
- 5.27 Direct flooding calculation ... 66
- 5.28 Stability after flooding ... 69
- 5.29 Floodable length ... 69
- 5.30 Subdivision criteria ... 70
    - 5.30.1 Factor of subdivision ... 71
    - 5.30.2 Compartmentation ... 71
    - 5.30.3 Probability of survival ... 71
- 5.31 References ... 71

6 Ships and the sea ... 73
- 6.1 Environmental elements ... 73
- 6.2 Waves ... 73
    - 6.2.1 Regular waves ... 74
    - 6.2.2 Irregular waves ... 76
    - 6.2.3 Wave energy ... 77
- 6.3 Ship response to the sea ... 78
    - 6.3.1 Oscillatory ship motions ... 78
    - 6.3.2 Rolling ... 79
    - 6.3.3 Pitching ... 79
    - 6.3.4 Heaving ... 79
    - 6.3.5 Cross coupling of motions ... 80

		6.3.6	Added mass and damping	80

	6.3.7	Motion in waves	80
6.4	Seaworthiness and seakeeping		82
	6.4.1	Strength	82
	6.4.2	Freeboard	82
	6.4.3	Stability	83
	6.4.4	Ship operation	83
	6.4.5	Seasickness	83
	6.4.6	Overall performance	83
	6.4.7	Motion control	83

7 Structural Strength 85

7.1	Longitudinal stresses		85
	7.1.1	Static loading	85
	7.1.2	Dynamic loading	87
7.2	Stressing of the ship's structure		88
7.3	Calculating static longitudinal stresses		88
	7.3.1	Characteristics of shear force and bending moment curves	90
7.4	Structural response		91
	7.4.1	Equivalent steel area	94
	7.4.2	Changing the section modulus	95
	7.4.3	Stresses when inclined	97
7.5	Local stresses		99
	7.5.1	Superstructures	100
7.6	Shear stresses		102
	7.6.1	Deflection of ships	102
7.7	Absolute and classification society stresses		104
7.8	Calculating dynamic longitudinal strength		107
	7.8.1	Strip theory	107
	7.8.2	Model tests	108
	7.8.3	Full-scale measurements	110
	7.8.4	A comparison of methods	110
7.9	Complex bending and torsion		111
7.10	Structural element strength		111
7.11	Calculating transverse strength		114
7.12	Finite element analysis		114
7.13	Structural failure and safety		115
7.14	References		116

8 Resistance 117

8.1	Ship and model correlation		117
8.2	Resistance components		120
	8.2.1	Frictional resistance	120
	8.2.2	Frictional resistance experiments	121
	8.2.3	ITTC ship-model correlation line	122
	8.2.4	Wavemaking resistance	125
	8.2.5	Other resistance effects	126
8.3	Resistance calculation		126

8.4	Correlating ship and model results	127
8.5	Standard series data	129
8.6	Full scale resistance testing	130
8.7	References	132

9 Propellers and Propulsion 133
9.1 Propellers 133
    9.1.1 Diameter 133
    9.1.2 Pitch 133
    9.1.3 Pitch ratio 135
    9.1.4 Blades 135
    9.1.5 Rake and skew 135
    9.1.6 Cap 135
9.2 Propeller action theories 135
    9.2.1 Momentum theory 136
    9.2.2 Blade element theory 138
9.3 Propeller and model correlation 141
9.4 Propeller coefficients 143
9.5 Propeller design 145
9.6 Propeller testing 147
    9.6.1 Propeller and ship interaction 147
9.7 Propeller design procedure 151
9.8 Model measurement of hull efficiency 151
9.9 Cavitation 152
    9.9.1 Ship and model cavitation correlation 152
    9.9.2 Cavitation number 152
    9.9.3 Types of cavitation 153
    9.9.4 Cavitation tunnel 154
9.10 Speed trials 154
    9.10.1 Ship speed measurement 154
    9.10.2 Analysis of trial data 155
9.11 Alternative propulsion systems 155
    9.11.1 Controllable pitch propellers 155
    9.11.2 Cycloidal Propellers 155
    9.11.3 Ducted propeller 156
    9.11.4 Contra-rotating propellers 157
    9.11.5 CLT propeller 157
    9.11.6 Grim Wheel 157
    9.11.7 Thrusters 157
    9.11.8 Waterjet 158
    9.11.9 Wind assistance 158

10 Manoeuvring and motion control 159
10.1 Directional control 159
    10.1.1 Directional stability 159
    10.1.2 The rudder 159
    10.1.3 Rudder operation and ship turning 162

          10.1.4 Assessing manoeuvrability............................165
    10.2 Types of rudder...167
          10.2.1 Conventional..167
          10.2.2 Special rudders and manoeuvring devices................167
    10.3 Motion control 169
          10.3.1 Bilge keels..170
          10.3.2 Fin stabilisers..171
          10.3.3 Tank stabilisers.......................................172
    10.4 Vertical lift control 173
          10.3.1 Hydrofoils..173

11 Vibration..175
    11.1 Basic concepts..175
          11.1.1 Damping effects 176
          11.1.2 Forced vibration 177
    11.2 Ship vibration 179
          11.2.1 Flexural vibrations of a beams........................180
          11.2.2 Ship flexural vibrations 181
          11.2.3 Ship vibration formulae 183
          11.2.4 Added mass 183
          11.2.5 Bending theory 184
    11.3 Approximate formulae 185
          11.3.1 Kumai's approach.......................................185
          11.3.2 Todd's approach 189
          11.3.3 Comparative approach 189
          11.3.4 Higher orders 191
    11.4 Direct calculation methods 192
          11.4.1 Deflection method 193
          11.4.2 Energy method 193
          11.4.3 Finite element method 194
    11.5 Noise 195
          11.5.1 Sound measurement 195
          11.5.2 Noise prediction 196
          11.5.3 Noise control 196
    11.6 References..197

12 Ship design 199
    12.1 Design objectives 199
    12.2 Design Category 201
    12.3 Design Process 202
          12.3.1 Concept design 203
          12.3.2 Preliminary Design.....................................205
          12.3.2 Contract Design 205
    12.4 Technology Impacts 206
    12.5 References...207

Index...209

1 Ship types

Merchant ships are designed to carry cargoes across the oceans of the world safely, speedily and economically. Specialist ships have been developed for some cargoes such as crude oil, bulk materials and wheeled vehicles. General cargo ships, on the other hand, will carry a variety of cargoes. Much of this general cargo is now carried in containers, or standard-sized boxes, and specialist ships of progressively increasing size have been developed to transport them.

Since a large part of the world's surface, approximately three-fifths, is covered by water, the merchant ship will doubtless exist, perhaps in forms not known today, for centuries to come. In travelling over the oceans of the world, the ship, its cargo and its crew, will be involved in many aspects of international life. The global nature of marine transportation, eg world-wide weather and climatic changes, availability of cargo handling facilities in ports and international regulations will, inevitably, influence the design of a ship.

Naval architecture is the science of ship design and extends to almost every type of floating structure. The large variety of subject areas within the discipline are leading to specialist studies in propulsion, resistance, strength of structures, etc. Each of these topics will be introduced and discussed, at a level appropriate to an introductory text of this nature.

A variety of problems arise when designing a ship to meet specific cargo-carrying requirements, and it is for the naval architect to produce the best possible solution. A prospective shipowner will state their requirements in terms of cargo, carrying capacity, speed, whether cargo-handling equipment to be carried or not, etc. The ship will be part of a transportation system, perhaps operating between a number of ports with different cargo-handling facilities. Time spent in port discharging and loading cargo must be minimised, since a ship is only earning money when it is transporting cargo. The nature and type of cargo-handling equipment, and also the layout of the cargo-carrying spaces, or holds, will be important in this respect.

In transporting its cargo safely, speedily and economically, the arrangement and strength of the ship's structure becomes important, as do the amount of buoyancy provided, the stability, ship resistance and mode of propulsion. The length, breadth and hull shape must provide sufficient buoyancy for the vessel to remain afloat. Various static and dynamic loadings are created on the floating structure, and it must be strong enough to withstand the many forces acting upon it. Stability is the ability of a ship, in still water, when moved by an external force, to return to an upright position, once the force is removed. The hull form must minimise resistance, in order to reduce both the amount of propulsive power needed and the fuel consumption.

Safety is an important concern in all aspects of ship design and operation, and the ship must be seaworthy. This term relates to an ability to remain afloat in all conditions of weather, the vessel's stability, and its behaviour in various sea states. The many construction and regulatory aspects of seaworthiness will be discussed in later chapters.

The regions of the world in which the vessel operates, its cruising range, and the climatic extremes encountered, are all factors in the design of a ship. Ocean-going vessels require tanks for fresh water and oil fuel storage. Stability and trim arrangements must be satisfactory in all weather states encountered. The strength of the structure, its ability to resist the effects of waves and heavy seas, etc, must all be greater for an ocean-going vessel than for a coastal or inland waterway vessel.

Many different types of merchant ship exist, and the list will increase as long as there is sufficient demand in a specialist trade. There are three principal types of ship in operation today: the general cargo carrier, the bulk cargo carrier and the passenger-carrying vessel. The general cargo carrier may be a general cargo ship, or a more specialised form, such as a containership, or a roll-on roll-off wheeled vehicle carrier. The bulk cargo carrier may be a tanker carrying liquids, or a dry bulk carrier carrying coal or iron ore. Passenger ships will include cruise liners and some of the larger ferries.

Variety also extends to the materials of construction and the choice of propulsion and other essential systems for this self-sufficient vehicle, which is also home to its crew and, sometimes, hundreds of passengers. Steel, aluminium, wood, and reinforced plastics are just some of the materials used in ship construction. Propulsion systems may be diesel engines, steam or gas turbines, with transmission systems that use mechanical, hydraulic or electric devices to drive the one or more propellers. The numerous systems which care for both the cargo and the crew may include refrigeration, heating, air conditioning, electricity generation and distribution, sewage treatment and so on.

A naval architect must consider all of the above aspects when designing a ship that will function safely, speedily and economically. An examination of the principal types of ship in existence today will serve as a useful introduction to the subject of naval architecture. The finished product will be reviewed and then the detailed elements of the ship design process will be progressively introduced in the chapters that follow.

The development of ship types over the years has been determined very largely by the nature of their cargoes. The various designs can, to some extent, be divided into general cargo, bulk cargo, and passenger vessels.

The general cargo carrier is a flexible design of vessel which will go anywhere and carry anything. Specialist forms of the general cargo carrier include containerships, roll-on roll-off ships and barge carriers. Bulk cargoes may be liquid, solid, or liquefied gas and particular designs of vessel exist for the carriage of each.

Passenger-carrying vessels include cruise liners and ferries. Many special types of vessel exist which perform particular functions or are developments of a particular aspect of technology. These include multi-hull vessels, such as catamarans, and hydrofoils and hovercraft.

Some of the more common ship types will now be examined in more detail.

1.1 General cargo ships

The general cargo ship has several large clear open cargo-carrying spaces or holds, see Fig 1.1. One or more separate decks may be present within the holds and are known as 'tween (in-between) decks. These provide increased flexibility when loading and unloading and permit cargo segregation as well as improved stability. Access to these holds is by openings in the deck known as hatchways.

Hatchways are made as large as strength considerations permit, in order to reduce the amount of horizontal movement of cargo within the ship. Hatch covers are made of steel sections, usually hinged together. They must be watertight and rest upon vertical plated structures, known as coamings, which surround the hatchway. The coamings of the upper or weather deck hatches extend some distance above the deck, in order to reduce the risk of flooding the holds in heavy seas.

Some form of cargo handling equipment, which may take the form of derricks and winches or, more usually, deck cranes is always provided. Deck cranes are fitted to many vessels, since they reduce cargo handling times and manpower require-

Merchant Ship Naval Architecture

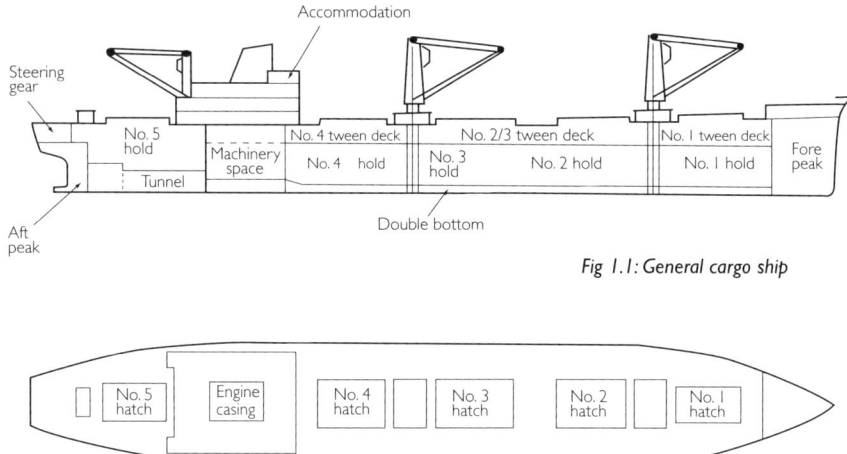

Fig 1.1: General cargo ship

ments. Some ships are equipped with a special heavy-lift derrick, which may serve one or two holds.

A double bottom is provided along the ship's length and is divided up into various tanks which may be used for fuel or lubricating oils, fresh water or sea water ballast. Fore and aft peak tanks are also fitted and may be used to carry liquid cargoes or water ballast. The water ballast tanks may be filled when the ship is only partially loaded, in order to provide a sufficient draught for stability and total propeller immersion to improve propulsion.

There is usually one hold aft of the accommodation and machinery space section of the vessel. This arrangement improves the trim of the vessel when it is partially loaded. The range of sizes for general cargo ships is from 2000 to 15 000 displacement tonnes with speeds from 12-18 knots.

1.2 Refrigerated cargo ships

The refrigerated cargo ship differs from the general cargo ship in that it carries perishable goods. A refrigeration system is necessary to provide low temperature holds for these cargoes, the holds and the various 'tween deck spaces being insulated to reduce heat transfer. The cargo may be carried frozen or chilled and various holds may be at different temperatures, according to the cargo requirements.

This type of vessel is usually faster than a general cargo ship, having speeds up to 22 knots. It is essentially a cargo liner, having set schedules and sailing between fixed terminal ports. Up to 12 passengers may be carried on some of these vessels. Displacements will range from about 15 000 to 20 000 tonnes.

1.3 Containerships

A container is a re-usable box of 2435mm wide and from 2435 to 2896mm high, with lengths of either 6055, 9125 or 12 190mm. Containers are now used for most general cargoes and liquid carrying versions – tank containers – also exist. Refrigerated containers are also in use which may have their own independent refrigeration plant or be supplied with suitable cold air from the ship's refrigeration system.

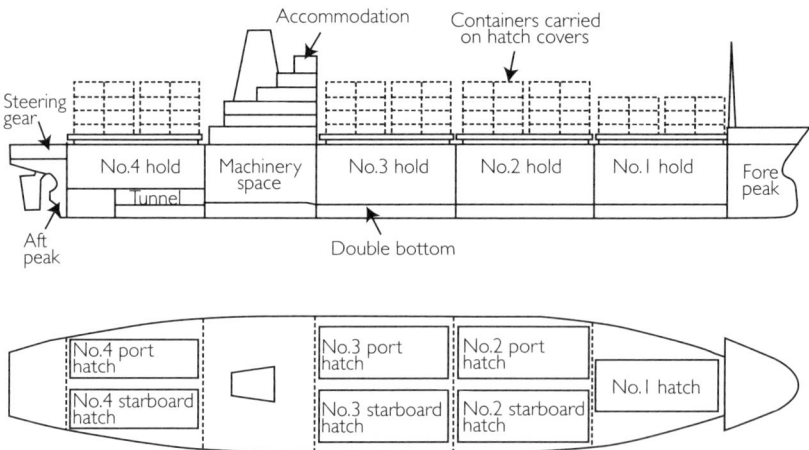

Fig 1.2: Containership

The cargo-carrying section of the containership is divided into several holds, each of which has a hatchway opening the full width and length of the hold, see Fig 1.2. The containers are racked in special frameworks and stacked one upon the other within the hold space. Cargo handling is thus only the vertical movement of the containers by a special quayside crane. Containers may also be stacked on the flush top hatch covers. Special lashing arrangements are used to secure this deck cargo.

The various cargo holds are separated by a deep web-framed structure to provide the ship with transverse strength. The ship structure outboard of the container holds on either side is a box-like arrangement of wing tanks which provides longitudinal strength to the structure. These wing tanks may be used for water ballast and can be arranged to counter the heeling of the ship when discharging containers. A double bottom is also fitted which adds to the longitudinal strength and provides additional ballast tank space.

The accommodation and machinery spaces are usually located aft to provide the maximum length of full-bodied ship for container stowage. Cargo-handling equipment is rarely fitted, except on some smaller vessels, since these ships travel between specially-equipped terminals to ensure rapid loading and discharge. Containership sizes vary considerably, with container carrying capacities from 1000 to 8000TEUs or more. The twenty-foot equivalent unit (TEU) represents a 20ft (6055mm) 'standard' container. Containerships are much faster than most cargo ships, with speeds up to 30 knots and operate as liners on set schedules between fixed ports.

1.4 Roll-on roll-off ships

This vessel was originally designed for wheeled cargo, usually in the form of trailers. The cargo could be rapidly loaded and unloaded via stern or bow ramps and sometimes sideports for smaller vehicles. The loss of cubic capacity due to undercarriages and clearances has resulted in many roll-on roll-off vessels also being adapted to carry containers on the deck.

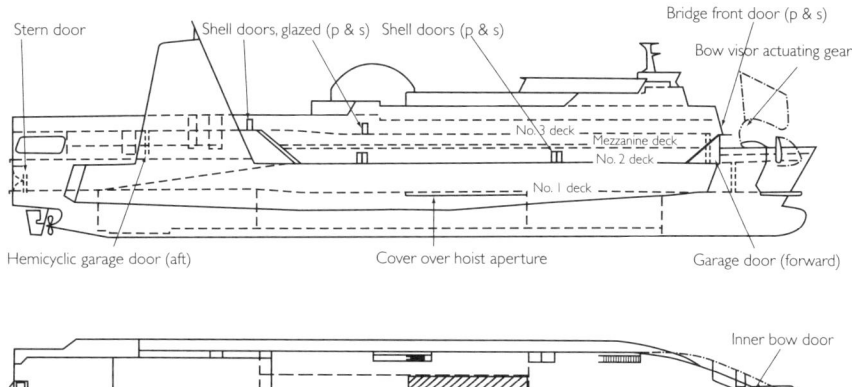
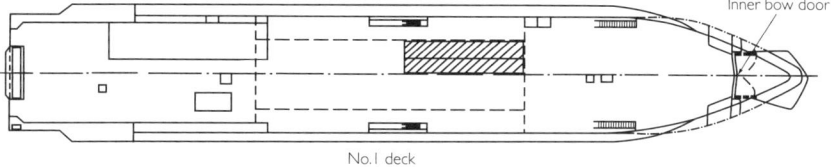

Fig 1.3: Roll-on and roll-off passenger ferry

The cargo carrying section of the ship is a large open deck with a loading ramp usually at the after end. Internal ramps lead from the loading deck to the other 'tween deck spaces. The cargo may be driven aboard under its own power, carried on trailers pulled by tractor units, or loaded by straddle carriers or fork lift trucks. One or more hatchways may be provided for containers or general cargo and will be served by one or more deck cranes. Arrangements may be made on deck for stowing containers. Some roll-on roll-off (ro-ro) vessels also have hatch covers to enable loading of the lower decks with containers. Where cargo (with or without wheels) is loaded and discharged by cranes, the term lift-on lift-off (lo-lo) is used.

The ship's structure outboard of the cargo decks is a box-like arrangement of wing tanks to provide longitudinal strength. A double bottom is also arranged along the complete length. The accommodation is located aft, as is the low-height machinery space. Only a narrow machinery-casing actually penetrates the loading deck. Sizes range considerably with about 16 000 deadweight tonnes (28 000 displacement tonnes) being quite common. High speeds in the region of 18-22 knots are usual. Many ro-ro ferries also carry passengers, Fig 1.3.

1.5 Barge carriers

This type of vessel is a variation of the containership. Instead of containers, standard barges are carried into which the cargo has been previously loaded. The barges, once unloaded, are towed away by tugs and return cargo barges are loaded. Minimal or even no port facilities are required and the system is particularly suited to countries with vast inland waterways. Two particular types will be described, the LASH (Lighter Aboard SHip), and the SEABEE.

The LASH ship carries barges, capable of holding up to 400 tonnes of cargo, which are 18.75m long by 9.5m beam and 3.96m deep. About 80 barges are carried stacked in holds much the same as containers, with some as deck cargo on top of the hatch covers. The barges are loaded and unloaded using a travelling gantry crane capable of lifting over 500 tonnes. Actual loading and discharging takes place between extended 'arms' at the after end of the ship. The ship structure around the barges is similar to a containership. The accommodation is located forward, whereas the machinery space is one hold-space forward of the stern. LASH ships are large, in the region of 45 000dwt (deadweight tonnes), with speeds in the region of 18 knots.

The SEABEE carries 38 barges, each of which may be loaded with up to 1000 tonnes of cargo. The barges have dimensions of 29.72m long by 10.67m beam and 3.81m depth and are loaded on board by an elevator located at the stern of the vessel. They are then winched forward along the various decks.

Deck hatchway openings do not exist in these vessels and the decks are sealed at the after end by large watertight doors. Two 'tween decks and the weather deck are used to store the barges. The machinery space and various bunker tanks are located beneath the 'tween decks.

The machinery space also extends into the box-like structure outboard of the barges to either side of the ship. The accommodation is also located here, together with several ballast tanks. A barge winch room is located forward of the barge decks and contains the machinery for horizontal movement of the barges. The SEABEE is physically about the same size as the LASH ship but with a slightly smaller deadweight of 38 000 tonnes. The speed is similarly in the region of 18 knots.

Despite being specialist vessels, both LASH and SEABEE can be used for other cargoes. Each can be used to carry containers and the SEABEE will also take ro-ro cargo. Other variations of barge carriers have been proposed such as the barge-carrying catamaran vessel (BACAT). Tug-barge systems have also been considered where the 'ship' is actually a number of linked barges with a separable propulsion unit.

1.6 Oil tankers

The demand for crude oil is constantly increasing. Oil tankers, in particular crude oil carriers, have significantly increased in size in order to obtain the economies of scale. Designations such as ULCC (Ultra Large Crude Carrier) and VLCC (Very Large Crude Carrier) have been used for these huge vessels. Crude oil tankers with deadweight tonnages in excess of half a million have been built, although the current trend is for somewhat smaller (100 000 – 150 000dwt) vessels. After the crude oil is refined the various products obtained, which include gas oil, aviation fuel and kerosene, are transported in products carriers.

The cargo-carrying section of the oil tanker is divided into individual tanks by longitudinal and transverse bulkheads. The size and location of these cargo tanks is dictated by the International Maritime Organization Convention MARPOL 1973/78. This 1973 Convention and its Protocol of 1978 further requires the use of segregated ballast tanks and their location such that they provide a barrier against accidental oil spillage. An oil tanker, when on a ballast voyage, must use only its segregated ballast tanks to achieve a safe operating condition. Double hulls are now a requirement for all new tankers.

One arrangement of a crude oil carrier that satisfies these requirements is shown in Fig 1.4. The cargo-carrying tanks include the centre tanks, wing tanks and two slop tanks. The segregated ballast tanks include all the double bottom tanks beneath the cargo tanks, the double hull and the fore and aft peak tanks. The cargo is discharged by cargo pumps fitted in the aft pump room. Each tank has its own suction arrangement that is connected, via a system of pipes and valves, to the cargo pumps which discharge the cargo ashore via further piping on deck and shore connections.

Considerable amounts of piping are visible on the deck running from the after pump room to the discharge manifolds positioned at midships, port and starboard.

Merchant Ship Naval Architecture

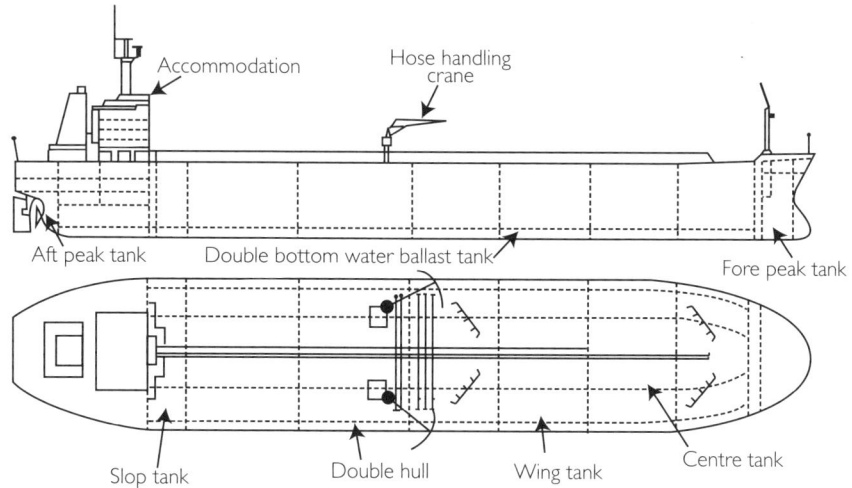

Fig 1.4: Oil tanker

Hose-handling derricks or cranes are fitted port and starboard near the manifolds. The accommodation and machinery spaces are located aft and separated from the tank region by a cofferdam, part of which may double as the pump room. The range of sizes for crude oil tankers is enormous, beginning at about 20 000dwt and extending beyond 500 000dwt. Speeds range from 12 to 16 knots.

The largest tankers are known as ULCCs and will be in excess of 300 000dwt, up to 362m long, with a breadth of 60m and a draught of about 22m. VLCCs are in the 250 000 to 300 000dwt range with length up to 330m, breadth of 58m and draught of about 20m. Neither vessel can pass through either the Panama or the Suez canals. A number of 'standard size' tankers are in use which meet certain trade or route restrictions, largely dictated by canals. Panamax vessels are able to pass through the Panama Canal and will be a maximum of 80 000dwt and not more than 305m long and 31m breadth. Suezmax are able to pass through the Suez Canal and will be a maximum of about 160 000dwt with a length up to 270m and a beam up to 48m. Aframax tankers are designed for African routes and the depths of these ports. They range from 97 000 to 107 000dwt, with a maximum length of 246m, a breadth of 42m and a draught of 14.5m.

Products carriers are oil tankers that carry the refined products of crude oil. The cargo tank arrangement is again dictated by MARPOL 73/78. Individual 'parcels' of various products may be carried at any one time, which results in several separate loading and discharging piping systems. The tank surfaces are usually coated to prevent cargo contamination and enable a high standard of tank cleanliness to be achieved after discharge. The current size range is from about 18 000 up to 75 000dwt, with speeds of about 14-16 knots.

1.7 Bulk carriers

The economies of scale have also been gained in the bulk carriage of solid cargoes such as grain, sugar and ore. A bulk carrier is a single-deck vessel with the cargo-carrying section of the ship divided into holds or tanks. The hold or tank arrangements vary according to the range of cargoes to be carried. Combination carriers are bulk carriers that have been designed to carry any one of several bulk cargoes on a particular voyage, eg, ore, crude oil or dry bulk cargo.

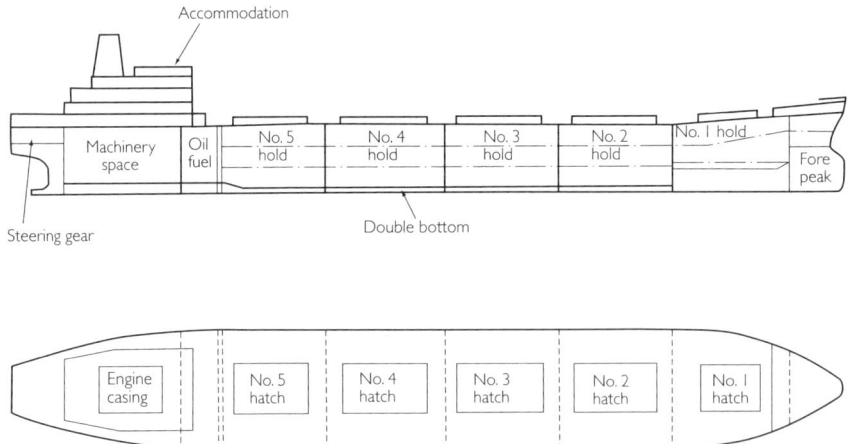

Fig 1.5: Bulk carrier

In a general-purpose bulk carrier, only the central section of the hold is used for cargo, see Figs 1.5 and 1.6(a). The partitioned tanks that surround the hold are used for ballast purposes when on empty or ballast voyages. The upper, or saddle, tanks may be ballasted in order to raise the ship's centre of gravity when a low-density cargo is carried. This hold shape also results in a self-trimming cargo. During unloading operations, the bulk cargo at the sides of the hold falls into the space below the hatchways and enables the use of grabs or other mechanical unloaders. Large hatchways are a particular feature of bulk carriers, since they reduce cargo-handling time during loading and unloading.

An ore carrier has two longitudinal bulkheads that divide the cargo section into wing tanks, port and starboard, and a centre hold which is used for ore. A deep double bottom is a particular feature of ore carriers. Ore, being a very dense cargo, would have a very low centre of gravity if placed in the hold of a normal ship. This would lead to an excess of stability in the fully loaded condition. The deep double bottom serves to raise the centre of gravity of the very dense cargo. The behaviour of the vessel at sea is thus much improved. On ballast voyages the wing tanks and the double bottoms provide ballast capacity. The ship cross-section would be similar to that for an ore/oil carrier shown in Fig 1.6 (b).

An ore/oil carrier uses two longitudinal bulkheads to divide the cargo section into centre and wing tanks which are used for the carriage of oil cargoes, see Fig 1.6 (b). When a cargo of ore is carried, only the centre tank section is used for cargo. A double bottom is fitted, but is only for water ballast. The bulkheads and hatch covers must be oiltight.

The ore/bulk/oil (OBO) bulk carrier is a popular combination vessel. It has a cargo-carrying cross-section similar to the general bulk carrier. The structure is, however, significantly stronger, since the bulkheads must be oiltight and the double bottom must withstand the concentrated, heavy ore load. Only the central tank or hold carries cargo, the other tank areas being ballast-only spaces, except the double bottoms, which may carry oil fuel or fresh water.

Large hatchways are a feature of all bulk carriers, in order to facilitate simple cargo handling. Many bulk carriers do not carry cargo-handling gear, since they trade between terminals which have special equipment. Where cargo handling equipment

Merchant Ship Naval Architecture

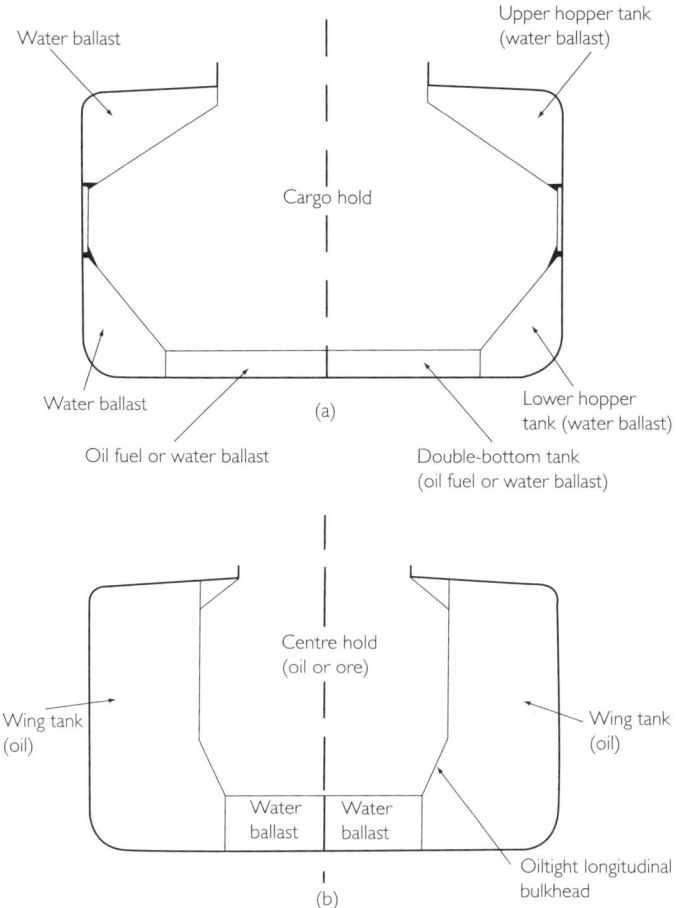

Fig: 1.6: Cross sections of the cargo areas of a standard type of bulk carrier (a) and of an ore-oil carrier (b)

is fitted (geared bulk carriers) this does make the vessel more flexible. Combination carriers handling oil cargoes have their own cargo pumps and piping systems for discharging oil. They will also be required to conform to the requirements of MARPOL 73/78. Their capacities range from small to upwards of 200 000dwt and they have speeds in the range of 12-16 knots.

A number of 'standard size' bulk carriers are in use which meet certain trade or route restrictions, largely dictated by canals. Capesize are bulk carriers too big for the Suez Canal, which travel around the Cape of Good Hope to Asia. Capesize vessels range from 80 000 to 199 000dwt with a beam of 40 to 50m. Handysize bulk carriers range from 10 000 to 35 000dwt and Handymax from 35 000 to 50 000dwt. Panamax vessels are able to pass through the Panama Canal and will be a maximum of 80 000dwt and not more than 305m long and 31m breadth. Suezmax are able to pass through the Suez Canal and will be a maximum of about 160 000dwt with a length up to 270m and beam up to 48m.

1.8 Liquefied gas carriers

The bulk transport of natural gases in liquefied form began in 1959, and has steadily increased since then. Specialist ships are now used to carry the various types of gases in a variety of tank systems, combined with arrangements for pressurising and refrigerating the gas.

Natural gas is found and released as a result of oil-drilling operations. It is a mixture of methane, ethane, propane, butane and pentane. The heavier gases, propane and butane, are termed 'petroleum gases'. The remainder, which consists largely of methane, is known as 'natural gas'. The properties and behaviour of these two basic gas groups vary considerably, thus requiring different means of containment and storage during transportation.

1.9 Liquefied natural gas carriers

Natural gas is, by proportion, 75-95% methane and has a boiling point of $-162°C$ at atmospheric pressure. Methane has a critical temperature of $-82°C$. This means that it cannot be liquefied by the application of pressure above this temperature. A pressure of 47 bar is necessary to liquefy methane at $-82°C$. It is not, therefore, possible to liquefy the gas at normal temperatures.

Liquefied natural gas carriers are designed to carry the gas in its liquid form at atmospheric pressure and a temperature in the region of $-164°C$. The ship design must, therefore, deal with protecting the steel structure from the low temperature, reducing the loss of gas, and avoiding its leakage into the other regions of the vessel.

Tank designs are divided into three main categories, namely self-supporting or free standing, membrane and semi-membrane. The self-supporting tank is constructed to accept any loads imposed by the cargo. A membrane tank requires the insulation between the tank and the hull to be load-bearing. Single or double metallic membranes may be used, with insulation separating the two membrane skins. The semi-membrane design has an almost rectangular cross-section and the tank is unsupported at the corners. Each arrangement uses a double hull type of construction with the space between the hulls being used for water ballast.

A liquefied natural gas carrier utilising semi-membrane tanks is shown in Fig 1.7. The cargo-carrying section is divided into five tanks of almost rectangular cross-section, each having a central dome. The liquid holding tank is made of 9% Ni steel while the secondary barrier is made of stainless steel. These two are supported and separated from the ship's structure by insulation, which is a lattice structure of wood and various foam compounds.

The tank and insulation structure is surrounded by a double hull. The double bottom and ship's side regions are used for oil or water ballast tanks, whilst the ends provide cofferdams between the cargo tanks. A pipe column is located at the centre of each tank and is used to route the pipes from the submerged cargo pumps out of the tank through the dome. The discharge piping is led along the decks to manifolds located near midships port and starboard. The accommodation and machinery spaces are located aft and separated from the tank region by a cofferdam. Liquefied natural gas carriers are being built in a large variety of sizes up to around 130 000m^3 (about 64 000dwt), and their speeds range from 16 to 19 knots.

1.10 Liquefied petroleum gas carriers

Petroleum gas may be propane, propylene, butane or a mixture of each. All three have critical temperatures above normal ambient temperature and can be liquefied at low temperature at atmospheric pressure, normal temperatures under considerable pressure, or some condition between. The ship design must, therefore, protect the steel hull where low temperatures are used, reduce the gas loss, avoid gas leakage and perhaps incorporate pressurised tanks.

Tank designs again divide into three main types, namely: fully pressurised; semi-

Merchant Ship Naval Architecture

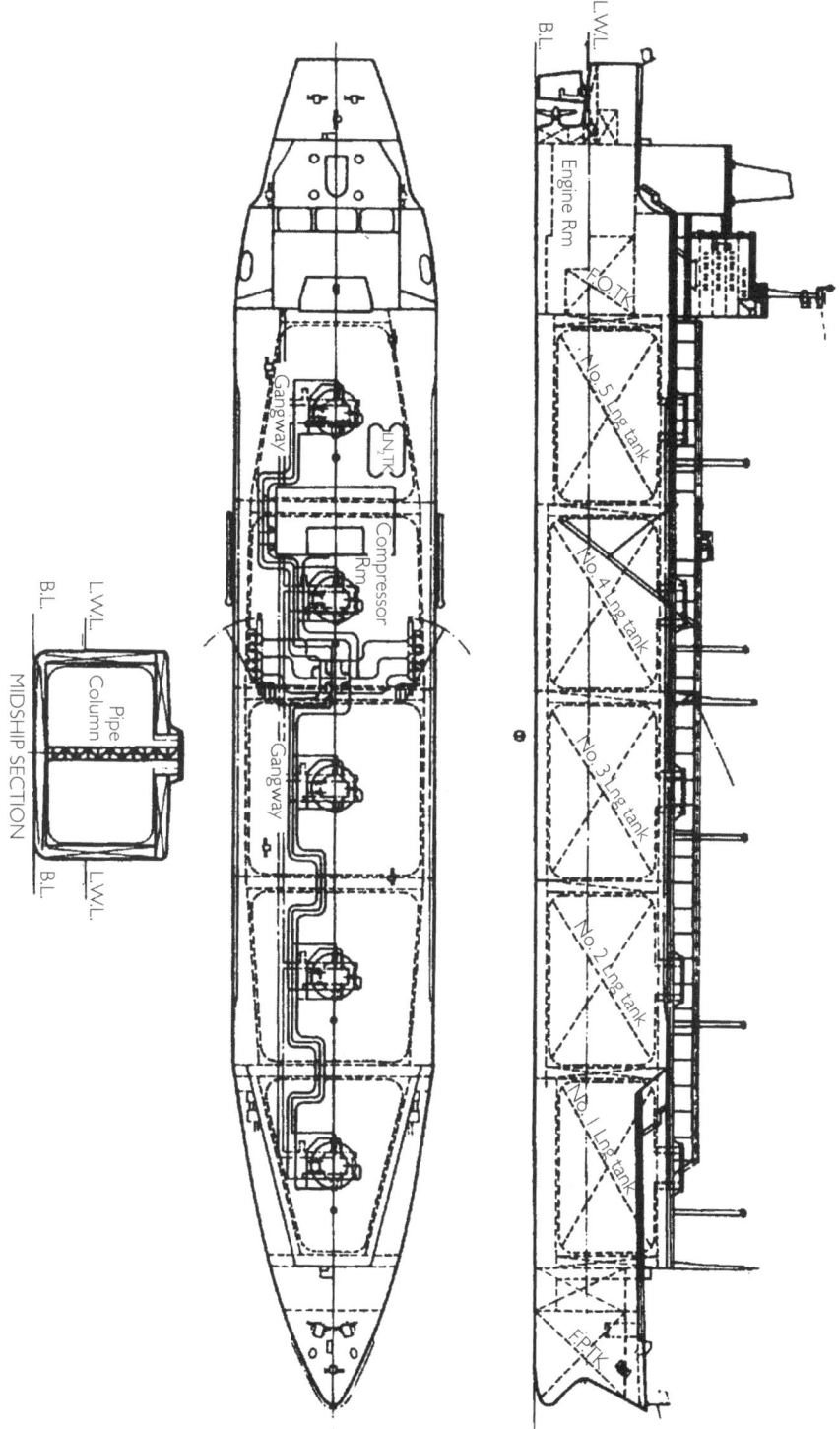

Fig 1.7: Liquefied natural gas carrier

pressurised, partially refrigerated; and fully refrigerated, atmospheric pressure arrangements. The fully pressurised tank operates at about 17 bar and is usually spherical or cylindrical in shape. The tanks usually protrude through the upper deck.

Semi-pressurised tanks operate at a pressure of about 8 bar and temperatures in the region of –7°C. Insulation is required around the tanks, and a reliquefaction plant is needed for the cargo boil-off. Cylindrical tanks are usual and they partially protrude through the deck. Fully refrigerated, atmospheric pressure, tank designs may be self-supporting, membrane or semi-membrane types, as previously described for liquefied natural gas tankers. The fully refrigerated tank designs operate at temperatures of about –45°C and a double hull type of construction is also used for this type of vessel.

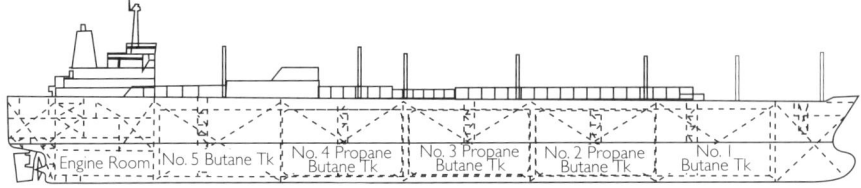

Fig 1.8: Liquefied petroleum gas carrier

A liquefied petroleum gas carrier utilising semi-membrane tanks is shown in Fig 1.8. The cargo carrying section is divided into five tanks. Tanks Nos 1 and 5 are used for the exclusive carriage of liquid butane, with the remainder being used for either butane or propane. The butane-only tanks are self-supporting, whereas the butane/propane tanks are of the semi-membrane type. The tank insulation in all cases uses polyurethane foam although the propane-carrying tanks also employ a lattice structure of wood. The propane tanks are refrigerated to about –45°C and the butane tanks to about –10°C. At these lower temperatures the inner hull is employed as the secondary barrier.

The double hull construction, cargo pumping arrangement, accommodation and machinery location are all similar to those of a liquefied natural gas carrier. A reliquefaction plant is, however, carried and any cargo boil-off is returned to the tanks after liquefying. Liquefied petroleum gas carriers are being built in sizes up to around 95 000m^3 (about 64 000dwt) and their speeds range from 16 to 19 knots.

1.11 Chemical tankers

A chemical tanker is a vessel constructed to carry liquids other than crude oil and products, or those cargoes requiring cooling or pressurised tanks. The vessel may carry chemicals, or even such liquids as wine, molasses or vegetable oils. Many of the chemical cargoes carried create a wide variety of hazards from reactivity, corrosivity, toxicity and flammability. Rules and regulations relating to their construction consider the effects these hazards have on the ship and it environment with respect to materials, structure, cargo containment and handling arrangements.

The IMO has produced a Code for the Construction and Equipment of Ships carrying Dangerous Chemicals in Bulk. This Code provides a basis for all such vessel designs, and the IMO Certificate of Fitness must be obtained from the Flag State Administration to indicate compliance. Also Annex II of MARPOL 73/78 Convention and Protocol is now in force and applies to hazardous liquid substances carried in chemical tankers.

Merchant Ship Naval Architecture

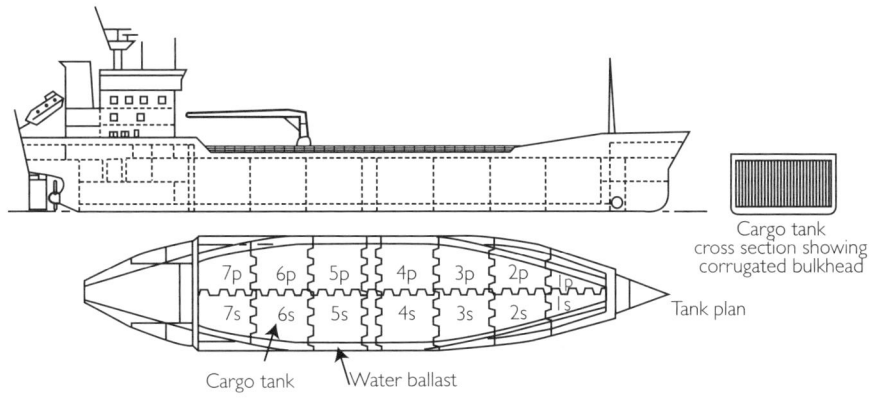

Fig 1.9: Chemical tanker

An IMO type II chemical tanker is shown in Fig 1.9. A double hull is used to protectively locate all the cargo tanks and even extends over the top. The cargo tank interiors are smooth, with all stiffeners and structure within the double hull. Corrugated bulkheads subdivide the cargo-carrying section into individual tanks. The double hull region of the double bottom and the ship's sides is arranged as water ballast tanks for ballast-only voyages or trimming and heeling when loaded.

Individual deepwell pumps are fitted in each cargo tank and also in the two slop tanks, which are positioned between tanks 4 and 5. Sizes for chemical tankers range from small coasters to vessels up to about 46 000dwt, with speeds of about 14-16 knots.

1.12 Passenger ships

Passenger ships can be considered in two categories, the luxury liner and the ocean-going ferry. The luxury liner is dedicated to the high-class transport of its human 'cargo'. The ocean-going ferry provides a necessary link in a transport system between countries and often carries roll-on roll-off cargo, in addition to its passengers.

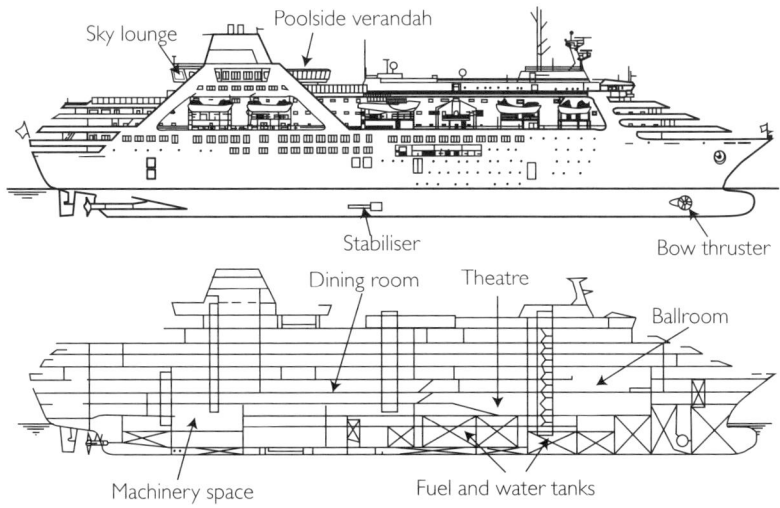

Fig 1.10: Passenger ship

Luxury passenger vessels are nowadays considered to be cruise liners, in that they provide high-class transport between interesting destinations in pleasant climates. The passenger is provided with a superior standard of accommodation and leisure facilities that results in large amounts of superstructure as a prominent feature of the vessel. The many tiers of decks are fitted with large open lounges, ballrooms, swimming pools and promenade areas. Aesthetically-pleasing hull lines are usual with well-raked clipper bows and unusual funnel shapes, see Fig 1.10. Stabilisers are fitted to reduce rolling and bow thrusters are used to improve manoeuvrability. Cruise liners range in size with passenger carrying capacities of around 2500 for a vessel of 91 000gt although several, larger, vessels of up to 150 000gt, with capacities for well over 3000 passengers, are in service. Speeds are usually high, in the region of 24 knots.

Ocean-going ferries are a combination of roll-on roll-off and passenger vessels. The vessel is therefore made up in three layers, the lower machinery space, the car decks and the passenger accommodation. A large stern door, and sometimes a lifting bow visor or sliding doors, provide access for the wheeled cargo. The stern door can also act as a ramp while a separate, weathertight inner door at the bow may be lowered to form a ramp for a shore connection. Within the vessel, fixed or movable ramps and in some instances large elevators, provide vehicular access between the various decks (see Fig 1.3).

The passenger accommodation will vary according to the length of the journey. For short-haul or channel crossings, public rooms with aircraft-type seats will be provided. On long distance ferries there are cabins and leisure facilities which may be up to the standard of cruise liners. Stabilisers and bow thrusters are also normally fitted to ocean-going ferries. Sizes will vary according to route requirements and speeds are usually high at around 20-22 knots.

1.13 Fast ferry designs

The current trend in ferry design is for fast, reliable vessels (> 25 knots), providing a high degree of passenger comfort, over relatively short sea routes (100 nautical miles). While ferry operators seek to meet passengers' needs, they also require vessels that are economical to purchase and operate, with efficient machinery which is cheap to maintain. Ferry designers and builders are meeting these, sometimes conflicting, needs in a variety of innovative ways, ranging from choice of materials to method of operation and choice of machinery and propulsion equipment. The final design of high speed ferries is inevitably a trade-off between speed and cost, performance and comfort, and size and bad weather performance.

The single or monohull design of vessel has been the traditional method of ship design which is still favoured for merchant ships. When used for fast ferries, a long slender hull with aesthetically-pleasing lines (hull shape) is usual and such vessels are available in a variety of designs from naval architects and shipbuilders worldwide. Construction is very cost effective but, operationally, the large area of hull in contact with the water increases resistance and requires large powerful machinery to propel it at speed. Its performance in bad weather is also poor, unless the vessel is very large.

Merchant Ship Naval Architecture

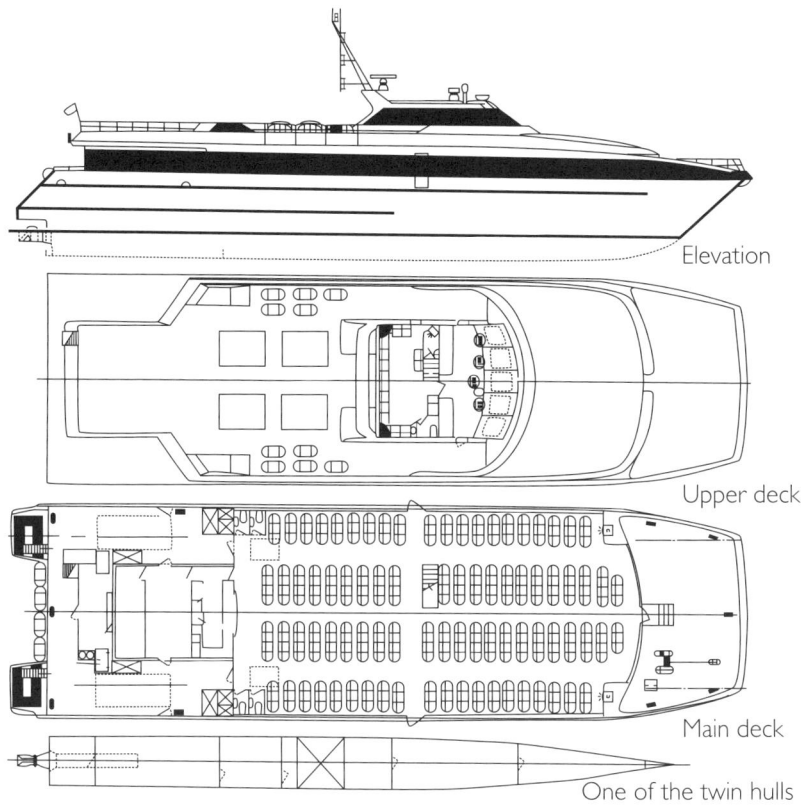

Fig 1.11: Passenger-carrying catamaran

Catamarans, or twin-hulled vessels, have become very popular for fast ferries over the last 30 or so years. The twin hull arrangement offers reduced resistance, enabling faster speeds, and a large platform for accommodation can be arranged between the two very narrow hulls, see Fig 1.11. The catamaran is more stable, can carry a large number of passengers and can be driven at high speed with reasonable engine power. The traditional design of catamaran does, however, lose speed and become uncomfortable in bad weather. The Wave Piercing Catamaran was an Australian-inspired development to combat the poor bad-weather performance of catamarans. Long narrow hulls are a feature of this design with knife-like leading edges to cut through oncoming waves, rather than ride up and down on them, thus improving the vessel's motion in rough seas.

The Hovercraft principle is to support the craft on a cushion of air and thus reduce its resistance, enabling high speeds for relatively-low engine power. Flexible skirts are used to retain the air under the vessel. While technically successful, the vessel has never really been commercially viable, due to its inability to operate in bad weather and considerable maintenance costs.

The Surface Effect Ship (SES), or side wall hovercraft, sought to combine the catamaran and the hovercraft by using catamaran hulls to enclose the air cushion on two sides and skirts at the fore and aft ends. The large payload of the catamaran was thus combined with the high speed of the hovercraft and, by careful monitoring of the air cushion pressure, some 'shock absorbing' is possible to improve passenger comfort.

Hydrofoil craft use foil or wing-like structures fore and aft to support the craft when at speed, utilising the same principle as an aeroplane wing. The fixed-foil hydrofoil provides high speed travel in comfort, using conventional propeller-driven systems that are low cost to purchase and maintain. Vessel performance is not good in bad weather. The Jetfoil craft took the development further, using fully submerged movable foils at either end of the craft. The expensive, complex, sophisticated foil control system produces what must be the most comfortable ferry ride even in poor weather. Operating costs and maintenance are expensive and the passenger payload is small.

Small Waterplane Area Twin Hull (SWATH) vessels have been developed to create a smooth platform in rough seas. The twin supporting, cigar-shaped hulls are positioned below the water surface, and the accommodation platform is supported on narrow struts that are not affected by surface wave action. Development continues with this vessel, but its operating costs are high because of the greater power needed to drive the large immersed hulls.

More recent developments in fast ferry design have sought to combine features of the designs outlined earlier. A foil-borne catamaran design operates at speeds up to 45 knots, when supported on foils mounted below the hulls. SLICE technology is a high speed ferry design of a SWATH vessel using four hulls instead of two, to reduce wake and enable higher speed.

1.14 Summary

This introduction, and brief examination of some ship designs in operation today, will set the scene for a more detailed study of the individual subject areas of naval architecture in subsequent chapters.

2 Defining a ship

Naval architecture has many terms which are unique, in order to precisely define and describe the features of a ship. The principal dimensions of a ship will be considered first, which describe its size, and then the terms used to describe important features and hull form.

2.1 Principal dimensions

The principal dimensions of length, breadth, depth and draught as they are measured for a ship will now be outlined and are shown in Fig 2.1.

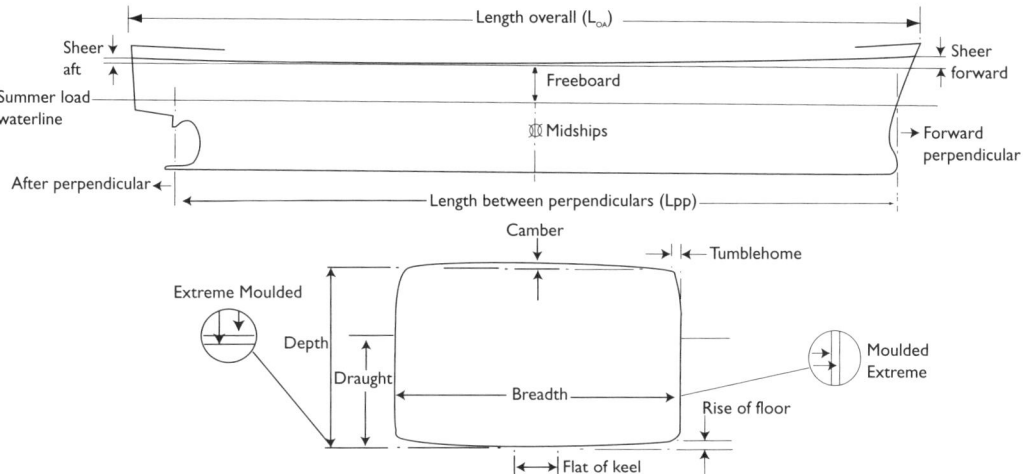

Fig 2.1: Ship dimensions and terminology

Perpendiculars are imaginary lines drawn on a ship to define points on the hull forward and aft. The **forward perpendicular** is a line drawn perpendicular to the waterline at the point where the forward edge of the stem intersects the summer load line. The **after perpendicular** is a line drawn perpendicular to the water line either (1) where the after edge of the rudder post meets the summer load line, or (2) in cases where no rudder post is fitted, the centreline of the rudder pintles (hinge pins) is taken.

The **length between perpendiculars** (L_{pp}) is the distance between the forward and after perpendiculars measured at the height of the summer load line. **Amidships** ⊗, or midships, is the point midway between the forward and after perpendiculars.

Length overall (L_{OA}) is the distance between the extreme points of the ship forward and aft.

The breadth, depth and draught of a ship may be measured from different datum points, resulting in measurements that are either extreme or moulded. **Extreme breadth** is the maximum measurement between the extreme points, port and starboard. **Moulded breadth** is the distance from port to starboard measured from the inside edges of the plating, ie it is the dimension for the sizes of the ship's frames. **Extreme depth** is the distance from the upper deck to the underside of the keel. **Extreme draught** is the distance from the waterline to the underside of the keel. The **base line**, or **moulded base line**, is a horizontal line drawn along the top edge of the keel from midships. **Moulded depth** is the distance from the upper deck to the **base line**, measured at the midship section. **Moulded draught** is the distance from the summer load line to the **base line**, measured at the midship section. The **draught** of a ship is the distance of the lowest point of the keel below the waterline.

2.2 Important features
Within the basic shape defined by the principal dimensions, a ship is further shaped by a number of important features, many of which have important practical applications, eg, camber to drain water from the decks, and flare to direct waves away from the forecastle (fo'c's'le).

Sheer is the curvature of the deck in a longitudinal direction. It is measured between the deck height at midships and the particular point along the deck.

Camber is the curvature of the deck in a transverse direction. It is measured between the deck height at the centre and the deck height at the ship's side.

Rise of floor is the height of bottom shell plating above the base line. Rise of floor is measured at the moulded beam line.

Bilge radius is the radius of the plating joining the side shell to the bottom shell. It is measured at midships.

Flat of keel is the width of the horizontal portion of the bottom shell, measured transversely.

Tumblehome is an inward curvature of the midship side shell in the region of the upper deck.

Flare is the outward curvature of the side shell at the forward end above the waterline.

Rake is a line in profile inclined from the vertical, indicating, for instance, the shape of the stem (the most forward part of the hull structure).

Freeboard is the vertical distance from the summer load waterline to the top of the freeboard deck plating, measured at the ship's side amidships. The uppermost complete deck exposed to the weather and sea is normally the freeboard deck. The freeboard deck must have permanent means of closure of all openings in it and below it. Minimum freeboard is a requirement under the Load Line Regulations, which are discussed in the next chapter.

Trim is the longitudinal inclination of a ship, determined by the difference between the forward and aft draughts. When the draughts are the same, the ship is said to be on an 'even keel'. 'Down by the head' indicates the draught forward is greater, and 'down by the stern' that the draught aft is greater.

2.3 Hull form
The complex form of a ship's hull is described in two-dimensional drawings known as the lines plan. This is a scale drawing of the moulded dimensions (to the inside of the plating) of the ship in plan, profile and section, see Fig 2.2.

The ship's length between forward and after perpendiculars is divided into ten equally-spaced divisions, or **stations**, numbered 0 to 10 beginning from aft. The after and forward perpendiculars are stations 0 and 10 respectively. Transverse sections of the ship at the various stations are drawn to produce a drawing known as the **body plan**. Since the vessel is symmetrical, half sections are given. The stations 0 to 5, representing the after half of the ship, are shown on the left side of the body plan with the forward section shown on the right.

The profile or **sheer plan** shows the general outline of the ship, any sheer of the decks, the deck positions and all the waterlines. The waterlines are a series of horizontal lines drawn at distances above the keel. For clarity, the deck positions have been omitted and only three waterlines are shown in Fig 2.2. The stations are also drawn on the sheer plan. Additional stations may be used at the fore and aft ends

Merchant Ship Naval Architecture

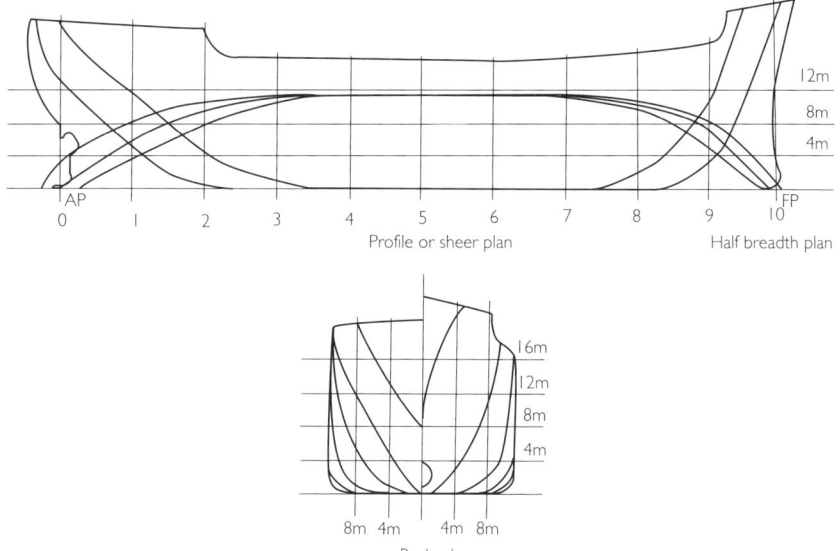

Fig 2.2: Lines plan

where the section change is considerable. The half breadth plan shows the shape of the waterlines and the decks formed by horizontal planes at the various waterline heights from the keel. The shapes formed at the various waterlines represent half waterplane areas. The **half breadth plan** is often superimposed on the sheer plan as shown in Fig 2.2.

The lines plan is drawn and then checked for '**fairness**'. To be 'fair' all the curved lines must run evenly and smoothly. There must also be exact correspondence between the dimensions shown for a particular point in the three different views. The fairing operation, once the exclusive province of a skilled loftsman, is now largely accomplished by computer programs.

The waterplane shape of a ship can be further described by a number of specialist terms. **Parallel middle body** is the length of the ship for which the midship section is constant in area and shape. **Entrance** is the immersed body of the ship forward of the parallel middle body, ie the fore body. **Run** is the immersed body of the ship aft of the parallel middle body, ie the after body.

A number of coefficients are used to describe the relationship between the ship's hull form and a surrounding regular shape. The **block coefficient**, C_B, is the ratio of the volume of displacement to a given waterline and the immersed volume of a constant rectangular section of the same length (L_{pp}), breadth and draught as the ship. The volume of displacement, ∇, is the volume of water occupied or displaced by a ship when floating in water.

Block coefficient, $\quad C_B = \dfrac{\nabla}{L_{PP} B T}$

where ∇ is the volume of displacement, L_{pp} is length between perpendiculars, B is extreme breadth and T is mean draught.

The **prismatic coefficient**, C_P, is the ratio of the volume of displacement to a given waterline to the immersed midship section and the length of the ship.

Prismatic coefficient, $C_P = \dfrac{\nabla}{L_{PP} A_M}$

where L_{PP} is the length between perpendiculars and A_M is the midship section area.

The **midship area coefficient**, C_M, is the ratio of the immersed are of the midship section to the area of a rectangle having the breadth of the ship and depth equal to the draught.

Midship area coefficient, $C_M = \dfrac{A_M}{BT}$

where A_M is the midship area, B is the breadth and T is the draught.

The above three coefficients are related as follows:
$$C_P = \dfrac{\nabla}{L_{PP} A_M} = \dfrac{\nabla}{L_{PP} BH C_M} = \dfrac{C_B}{C_M}$$

or $C_B = C_P \times C_M$

The **waterplane area coefficient** C_W, is the ratio of the area of the waterplane to the area of the rectangle equal to the length and breadth of the ship.

Waterplane area coefficient, $C_W = \dfrac{A_W}{L_{PP} \times B}$

The **vertical prismatic coefficient** C_{PV} is the ratio of the immersed volume to the area of the load waterplane and the mean draught.

Vertical prismatic coefficient, $C_{PV} = \dfrac{\nabla}{A_W \times T}$

$$= \dfrac{C_B}{C_W}$$

These coefficients are useful indicators of a ship's form in the early stages of the design. A range of values of these coefficients for different vessels is given in Table 2.1. A block coefficient approaching 1 will indicate a 'full' form shape whereas a value of 0.7 would be a 'fine' form. A low value of midship section coefficient will indicate well rounded bilges and a high rise of floor. A large value of vertical prismatic coefficient will indicate the body sections are U-shaped.

Ship type	C_B	C_P	C_M
General cargo ship	0.68-0.78	0.62-0.82	0.95-0.98
Passenger ship	0.57-0.72	0.64-0.77	0.95-0.96
Oil tanker	0.75-0.80	0.70-84	0.98-0.99
Bulk carrier	0.73-0.84	0.68-0.83	0.96-0.98

Table 2.1 Typical hull form coefficients

2.4 Displacement

The **displacement** of a ship refers to the weight of water displaced by a floating ship. It is the weight of the ship and will be a force expressed in newtons. Displacement is, however, generally referred to as a mass and expressed in tonnes.

Displacement = $\rho g \nabla$ newtons,

where ρ is the density of the sea water in which the ship is floating, g is acceleration due to gravity and ∇ is the volume of displacement.

The **lightweight** of a ship is the displacement when complete and ready for sea, but without crew, passengers, stores, fuel or cargo on board, ie the hull and machinery. The **deadweight** is the difference between the displacement and the lightweight at any given draught. Deadweight is the cargo, fuel, stores, etc, that a ship can carry. Cargo deadweight refers to the cargo alone. All these deadweight values are generally given as masses, in tonnes.

2.5 Tonnage

Tonnage is a measure of the internal capacity of a ship where 100ft^3 or 2.82m^3 represents 1 ton. Tonnage is a volume measurement, despite the somewhat confusing term 'ton' being used. Two values are used, **gross tonnage** which is the total internal capacity of the ship and **net tonnage** is the revenue-earning capacity after some deductions. The calculation of tonnage is described in the next chapter. Tonnage values are also used to determine port and canal dues, safety equipment and manning requirements and are a statistical basis for measuring the size of a country's merchant fleet.

3 Rules, regulations and legislation

The design, construction and safe operation of merchant ships is considerably influenced and regulated by a number of organisations and their various requirements. These organisations fall into three broad categories; classification societies, national and international authorities.

Classification societies, with their rules and regulations relating to ship classification, provide a set of standards for sound merchant ship construction which have been developed over many years. These 'rules' are based on experience, practical knowledge and considerable research and investigation.

Much of the legislation applied to ships is usually administered and enforced by the appropriate government department of the country where the ship is registered (Flag State). The load line rules and tonnage measurement are two particular legislative requirements which will be outlined.

The International Maritime Organization (IMO) is a specialist agency of the United Nations and is seeking to develop high standards in every respect of ship construction and operation. It is intended, ultimately, to apply these standards internationally to every merchant ship at sea.

3.1 Classification societies

A classification society exists to classify or 'arrange in order of merit' ships that are built according to its rules or are offered for classification. A classed ship is considered to have a particular standard of seaworthiness. There are classification societies within most of the major maritime nations of the world and some are listed below.

American Bureau of Shipping (USA) (www.eagle.org)
Bureau Veritas (France) (www.bureauveritas.com)
China Classification Society (China) (www.ccs.org.cn)
Det Norske Veritas (Norway) (www.dnv.com)
Germanischer Lloyd (Germany) (www.gl-group.com)
Korean Classification Society (Korea) (www.krs.co.kr)
Lloyd's Register (UK) (www.lr.org)
Nippon Kaiji Kyokai (Japan) (www.classnk.or.jp)
Register of Shipping (Russia) (www.rs-head.spb.ru)
Registro Italiano Navale International (Italy) (www.rina.org)

Consultation between the societies takes place on matters of common interest through the International Association of Classification Societies (IACS).

Classification societies operate by formulating and publishing rules and regulations relating to the ships' structural integrity and the reliability of their propelling machinery and equipment. These rules are the result of years of experience, research and investigation into ship design and construction. They are effectively a set of standards. There is no legal requirement for a shipowner to have a ship classified. However, insurance premiums paid may depend upon the class of a ship; the higher the standard, the lower the premium. Also, by being classified, a ship is shown to be of sound construction and a safe means of transport for cargo or passengers. There is no connection between the insurance companies and the classification societies.

The operation of Lloyd's Register, the oldest classification society, will now be considered. Throughout this book, all references to classification society rules are to those of Lloyd's Register. This society is run by a General Committee composed of members of the world's community and the industry which it serves. National committees are formed in many countries for liaison purposes. A Technical Committee advises the General Committee on technical problems connected with the society's business and any proposed alteration in the rules. The society publishes its *Rules and Regulations for the Classification of Ships* in book and CD form, which is updated as necessary, also *Extracts* from these rules and *Guidance Notes* relating to more specific structures and equipment. The society employs surveyors who ensure compliance with the rules by attendance during construction, repairs and maintenance throughout the life of classed ships.

To be classed with Lloyd's Register, approval is necessary for the construction plans, the materials used, and the construction methods and standards as observed by the surveyor. The rules governing the scantlings (sizes) of the ship's structure have been developed from theoretical and empirical considerations. Lloyd's Register collects information on the nature and cause of all ship casualties. Analysis of this information often results in modification to the rules to produce a structure which is considered to be adequate. Much research and investigation is also carried out by the society, leading likewise to modification and amendments to the rules.

The assigning of a class then follows acceptance by the General Committee of the surveyor's report on the ship. The highest class awarded by Lloyd's is: ✠100 A1. This is made up as follows:

100A   refers to the hull, when built to the highest standards laid down in the rules
1   refers to the equipment, such as the anchors and cables, being in good and efficient condition
✠   indicates the vessel has been built under the supervision of the society's surveyors.

It is also usual to name the type of ship followed by the classification, eg, ✠100A1 Oil Tanker. Machinery is also surveyed and the notation LMC (Lloyd's Machinery Certificate) is used, where the machinery has been built according to the society's rules and satisfactorily proved on sea trials. This information regarding the classification of a ship is entered in the *Register of Ships*. The *Register of Ships* is a book containing all the names, classes and general information concerning the ships classed by Lloyd's Register, and also particulars of all known ocean going merchant ships in the world of 100gt (gross tonnage is a capacity measure – see Chapter 2, Tonnage) and upwards.

The maintaining of standards is ensured by the society requiring all vessels to have annual surveys or examinations. Special surveys are also required every four years from the date of the first survey for classification. The society is also empowered to act as an assigning authority. This means that it acts as the agent for the government in administering some of the mandatory requirements for shipping, eg, the Load Line Rules.

3.2   National authorities

Legislation relating to the safe operation of ships is the responsibility of the government of the country in which the ship is registered, ie, the Flag State. In the United Kingdom this is the concern of the Maritime and Coastguard Agency (MCA), an executive arm of the Department of the Environment, Transport and the Regions. This Agency is empowered to draw up rules by virtue of a number of Merchant Shipping Acts extending back more than a hundred years. The MCA employs surveyors who examine ships to verify that they are built in accordance with the regulations. Some of the matters with which the MCA is concerned are:

- Load lines.
- Tonnage.
- Master and crew spaces.
- Watertight subdivision of passenger ships.
- Life-saving appliances.

- Carriage of grain cargoes.
- Dangerous cargoes.

Several of these topics are the subject of international regulations, eg, load lines, tonnage and regulations related to passenger ships. Load lines and tonnage will be considered in more detail later in this chapter, as they significantly affect ship design.

3.3 IMO

The International Maritime Organization (IMO) is a specialist agency of the United Nations and was originally set up to enable intergovernmental co-operation on matters concerning ships, shipping and the sea. The first Assembly took place in 1959. The Assembly is the governing body of IMO, which meets once every two years and consists of all the member states. IMO has 165 members and three associate members, Hong Kong, Macau and Faroe Islands.

About 60 organisations have Non-Governmental Organisation status with IMO and can attend meetings, present papers and speak, but not vote. Up-to-date information on IMO can be found on its home page (http://www.imo.org). Between sessions of the Assembly, the affairs of the organisation are run by the Council, which consists of 40 member governments elected for two-year terms by the Assembly. The organisation's technical work is carried out by a number of committees, the most senior of which is the Maritime Safety Committee (MSC). This has a number of sub-committees whose titles indicate their subject specialism. They are the sub-committees on Safety of Navigation; Radio Communications and Search and Rescue; Training and Watchkeeping; Carriage of Dangerous Goods, Solid Cargoes and Containers; Ship Design and Equipment; Fire Protection; Stability and Load Lines and Fishing Vessel Safety; Flag State Implementation; and Bulk Liquids and Gases.

As a result of the legal issues involved in much of its work, the organisation also has a Legal Committee, and a Technical Co-operation Committee which co-ordinates and directs IMO's activities in the provision of technical assistance in the maritime field particularly to developing countries. The Facilitation Committee deals with measures to simplify and minimise documentation in international maritime traffic. The aim is to reduce the formalities and simplify documentation needed by ships when entering or leaving ports.

In order to achieve its objectives, IMO had, by the year 2004, promoted the adoption of over 40 conventions and protocols. It has also adopted well over 800 codes and recommendations on various matters relating to maritime safety and the prevention of pollution. The initial work on a convention is normally done by a committee or sub-committee. A draft instrument is then produced which delegations from all states with the United Nations, including states which may not be IMO members, are invited to comment. The Conference adopts a final text, which is submitted to IMO member governments for ratification.

An instrument so adopted comes into force after fulfilling certain requirements, which usually include ratification by a specified number of countries whose merchant fleet comprises a certain percentage of the world fleet. Generally speaking, the more important the convention, the more stringent are the requirements for entry into force. Observance of the requirements of a convention is mandatory for countries that are parties to it. Codes and recommendations which are adopted by the IMO Assembly are not so binding on governments. However, their contents can be just as important, and in most cases they are implemented by governments by incorporation into their legislation.

Conventions must be kept up-to-date by amendments, which are formally adopted at an IMO Conference. The initial procedure for contracting governments' acceptance was an explicit acceptance by a proportion, usually two-thirds. This process has proved too slow and now a 'tacit acceptance' procedure is used. Under this system, the amendment enters into force on a date selected by the conference or meeting at which it is adopted, unless it is rejected by a specified number of contracting parties, usually one-third.

Safety and the prevention of pollution are the two chief concerns of IMO and the work done in these areas which has an influence on ship design, will now be considered.

3.4 Safety

The first conference organised by IMO in 1960 was concerned with maritime safety and adopted the **International Convention on Safety of Life at Sea (SOLAS)**, which came into force in 1965, replacing a version adopted in 1948. The 1960 SOLAS Convention covered a wide range of measures designed to improve the safety of shipping. The 1960 Convention was amended several times. However, because of the difficult requirements for bringing amendments into force, none of these amendments actually became binding internationally. To remedy this situation and introduce needed improvements more speedily, IMO adopted a new version of SOLAS in 1974, which incorporated the amendments adopted to the 1960 Convention as well as other changes, including an improved amendment procedure. Under the new procedure, amendments adopted by the MSC would enter into force on a predetermined date, unless they were objected to by a specific number of states. The 1974 SOLAS Convention entered into force on 25 May 1980, since when it has been modified on a number of occasions.

Additionally, two protocols have been adopted to the Convention: the 1978 Protocol, which modified inspection and survey procedures and introduced mandatory annual surveys and inspections for tankers (in force since 1984); and the 1988 Protocol, which introduced a harmonised system of survey and certification, among other things (in force since 2000).

The various chapters of SOLAS 1974 deal with vessel construction in relation to subdivision and stability; machinery and electrical installations; fire protection, detection and extinction; life-saving appliances; radio communications; safety of navigation; carriage of various cargoes; nuclear ships; high speed craft and special measures to enhance maritime safety. Appendices deal with certificates and some amendments.

In 1966 a conference convened by IMO adopted the **International Convention on Load Lines**. Limitations on the draught to which a ship may be loaded, in the form of freeboards, are an important contribution to its safety. An international convention on the subject had been adopted in 1930 and the new instrument brought this up to date and incorporated new and improved measures. It came into force in 1968 and has been amended several times since then. A Protocol of 1988 contains modifications agreed at the International Conference on the Harmonised System of Survey and Certification.

The system of tonnage measurement of ships can also affect safety and this has been one of the most difficult problems in all maritime legislation. Tonnage is used for assessing dues and taxes and, because of the way in which it is calculated, it has been possible to manipulate the design of ships to reduce the ship's tonnage, while still allowing it to carry the same amount of cargo. This has, however, sometimes been at the expense of the vessel's safety and stability. Several systems of tonnage measurement were developed over the years, but none was universally recognised. IMO began work on this subject soon after coming into being, and in 1969 the **International Convention on Tonnage Measurement of Ships** was adopted. It is an indication of the complexity of the matter that the Convention did not enter into force until 1982.

IMO's work involves more than just the adoption of conventions. The organisation has produced numerous codes, recommendations and other instruments dealing with safety. These do not have the same legal power as conventions, but can be used by governments as a basis for domestic legislation or guidance. Topics covered have included bulk cargoes, the carriage of dangerous goods, the carriage of bulk chemicals, liquefied gases, noise levels on board ships, and special purpose ships. Some codes have been incorporated into conventions as amendments.

3.5 Prevention of pollution

The 1954 **Oil Pollution Convention** was the first major convention designed to curb the impact of oil pollution. The 1954 Convention was amended in 1962, but the wreck of the tanker *Torrey Canyon* in 1967 alerted the world to the great dangers which the transport of oil poses to the marine environment. Following this disaster, IMO produced a series of conventions and other instruments, including further amendments to the 1954 Convention which were adopted in 1969. In 1973 IMO convened a major conference to discuss the whole problem of marine pollution from ships. It resulted in the adoption of the first-ever comprehensive anti-pollution convention, the **International Convention for the Prevention of Pollution from Ships (MARPOL)**.

The convention deals not only with pollution by oil, but also pollution from chemicals, other harmful substances, garbage and sewage. The MARPOL Convention greatly reduces the amount of oil that may be discharged into the sea by ships, and completely bans such discharges in certain areas (such as the Black Sea, Red Sea and other regions). Certain technical problems made it difficult for many States to ratify the convention, and a series of tanker accidents in the winter of 1976/77 led to demands for further action. IMO convened the Conference on Tanker Safety and Pollution Prevention in 1978. This Conference adopted a protocol to the 1973 MARPOL Convention which introduced further measures, including requirements for such operational techniques as crude oil washing (a development of the earlier 'load on top' system) and a number of modified constructional requirements such as protectively located segregated ballast tanks. The Protocol of 1978 relating to the 1973 MARPOL Convention in effect absorbs the parent Convention with modifications. This combined instrument is commonly referred to as MARPOL 73/78 and entered into force in October 1983. The convention has been amended on several occasions since then.

The design of oil tankers is heavily influenced by MARPOL 73/78 requirements with tank sizes limited, requirements for separate ballast-only tanks and double bottoms and wing tanks extending over the full depth of the ship's side. All new oil tankers must now have a double hull throughout the cargo tank area.

3.6 Fire safety in ships

Ship design and construction is also influenced by the arrangements for fire protection as detailed in the 1974 International Convention on Safety of Life at Sea (SOLAS 74) and also classification society rules. These rules are particularly stringent for passenger ships carrying more than 36 passengers, and cargo ships of more than 4000gt.

The following principles are the basis of the regulations:
(1) The use of thermal and structural boundaries to divide the ship into main vertical fire zones.
(2) Thermal and structural boundaries are used to separate the accommodation spaces from the rest of the ship.
(3) The use of combustible material is to be restricted.
(4) Any fire should be detected, contained and extinguished where it occurs.
(5) Access must be provided to enable fire-fighting and a protected means of escape.
(6) Where inflammable cargo vapour exists, the possibility of its ignition must be minimised.

The above arrangements are made to ensure that a fire on board a ship will be contained within the zone in which it occurs. Attempts can then be made to extinguish the fire or, at worst, escape.

3.7 The load line rules – freeboard

Freeboard is the distance measured from the waterline to the upper edge of the deck plating at the side of the freeboard deck amidships. The load line rules set out the requirements for a minimum freeboard, which must be indicated on the ship's side by a special load line mark. This minimum freeboard is a statutory requirement in the United Kingdom under the Merchant Shipping (Load line) Rules of 1968. These rules are based on the **International Convention on Loadlines, 1966**.

A minimum freeboard is required to ensure that the ship is seaworthy when loaded. The minimum freeboard provides the ship with a reserve of buoyancy which enables it to rise as it passes through waves and thus remain largely dry on its decks. This reserve buoyancy also improves the vessel's stability and, in the event of damage, will enable it to remain afloat indefinitely, or at least for a time to enable the crew to escape.

The assigning of freeboard follows a calculation which considers the ship's length, breadth, depth and sheer, the density of the water and the amount of watertight superstructures and other features of the ship. Additional conditions of assignment are also made in relation to certain openings and fittings. The ship is assigned a basic minimum freeboard on the assumption that it is correctly loaded, with adequate stability and strength.

In order to assign freeboards, ships are divided into Types A and B. Type A ships are those designed specifically for the carriage of liquid cargoes in bulk. Type B are all other ships. Type A ships have a smaller freeboard on the basis of having only small openings for access, which are covered by watertight covers of adequate strength. A ship's freeboard is established from a calculation where a tabular freeboard figure, based on the ship's length and type, is adjusted by several corrections, such as sheer, depth and superstructure.

The maximum summer draught, as determined from the load line calculations is indicated by a **load line mark**. This consists of a ring of 300mm outside diameter and 25mm wide, intersected by a horizontal line 450mm long and 25mm wide. The upper edge of this line passes through the centre of the ring. The ring is positioned at midships and at a distance below the upper edge of the deck line that corresponds to the assigned minimum summer freeboard. This value may not be less than 50mm. The **deck line** is a horizontal line, 300mm long and 25mm wide, which is positioned amidships port and starboard. The upper edge of the line is located on the outer shell level with the upper surface of the freeboard deck plating.

A series of load lines are situated forward of the load line mark and these denote the minimum freeboards within certain geographical zones, or in fresh water. The summer load line is level with the centre of the ring and is marked S. The tropical, T, and winter, W, load lines are found by deducting and adding – from and to the summer freeboard respectively – 1/48 of the vessel's summer moulded draught. For a ship of 100m or less, a Winter North Atlantic (WNA) zone load line is permitted. This line is positioned at the winter freeboard plus 50mm. The fresh water freeboards, F, and TF are determined by deducting from the summer or tropical freeboard the value:

$$\frac{\text{Displacement in salt water}}{4 \times \text{TPC}} \text{ millimetres,}$$

where TPC is the tonnes per centimetre immersion in salt water at the summer load waterline.

These markings are shown in Fig 3.1. In all cases, measurements are to the upper edge of the line.

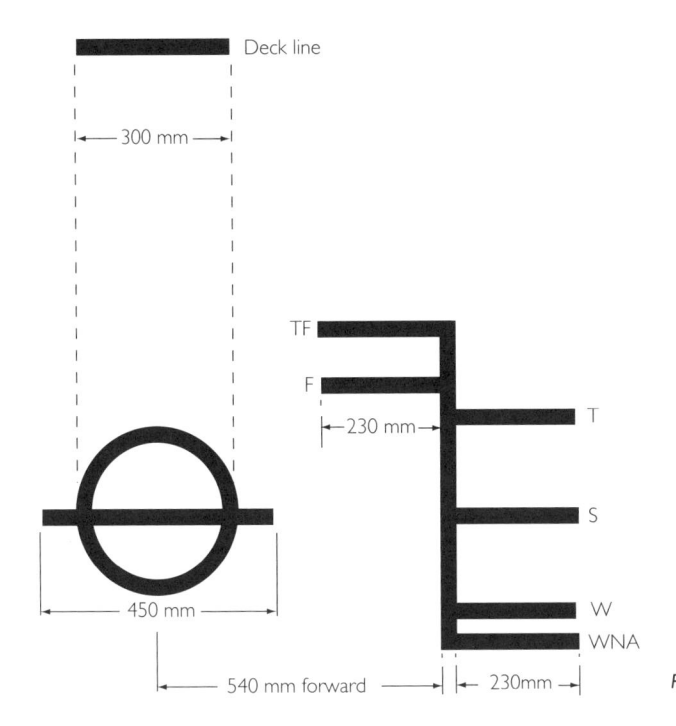

Fig 3.1: Load line markings

3.8 Tonnage

Tonnage, as discussed in this section, is a measure of cubic capacity where 1 ton represents 100ft³ or 2.83m³. Tonnage is a measure of the ship's internal capacity, with two values being used. The **gross tonnage** is the total internal capacity of the ship and the **net tonnage** is the revenue-earning capacity. Tonnage values are also used to determine port and canal dues, safety equipment and manning requirements and are a statistical basis for measuring the size of a country's merchant fleet. All ships, prior to being registered, must be measured according to their country's tonnage regulations. The differences in the various measuring systems have led to ships having several tonnage values and to unusual designs, which exploited aspects of tonnage measurement. The 1969 IMO International Conference on Tonnage Measurement of Ships led to an international review of the subject and a system which has now been universally adopted.

3.9 The International Convention on Tonnage Measurement of Ships

This Convention, which was the first successful attempt to create a universal tonnage measurement system, came into force on 18 July 1982. All aspects of the Convention became fully operative on 18 July 1994.

Gross and net tonnage are the only two parameters now used. Gross tonnage is determined in relation to the volume of all enclosed spaces. Net tonnage is the sum of the cargo space plus any volume for passenger spaces multiplied by a coefficient to bring the value close to tonnages of earlier schemes. Each measurement is determined by a formula as follows:

$$\text{Gross tonnage}\,(GT) = K_1 V \qquad \text{Net tonnage}\,(NT) = K_2 V_c \left(\frac{4d}{3D}\right)^2 + K_3\left(N1 + \frac{N2}{10}\right)$$

where
V = total volume of all enclosed spaces in the ship in cubic metres
$K_1 = 0.2 + 0.02 \log_{10} V$
V_c = total volume of cargo spaces in m³
$K_2 = 0.2 + 0.02 \log_{10} V_c$ [handwritten: VIDE AMMENDMENT NOTED AT LAST PAGE]
$K_3 = 1.25 \dfrac{GT + 10\,000}{10\,000}$

D = moulded depth amidships in m
d = moulded draught amidships in m
N_1 = number of passengers in cabins with not more than eight berths
N_2 = number of other passengers
$N_1 + N_2$ = total number of passengers the ship is permitted to carry, as indicated on the ship's passenger certificate. When $N_1 + N_2$ is less than 13, N_1 and N_2 shall be taken as zero.
GT = gross tonnage of the ship.

In the above calculations, the factor $(4d/3D)^2$ is not to be taken as greater than unity and the term $K_2 V_c (4d/3D)^2$ is not to be taken as less than 0.25GT.

The volumes referred to in these formulae are to be calculated to the inside of plating and include the volumes of appendages. Volumes of spaces open to the sea are excluded.

The main features of this convention can be summarised as follows:
1. Measurements of gross and net tonnage are dimensionless numbers. The word ton will no longer be used.
2. New ships were defined as ships whose keels were laid, or were at a similar stage of construction, on or after 18 July 1982.
3. Existing ships were allowed to retain their then current tonnages until 18 July, 1994. After that date they could retain their existing tonnages only for the purpose of the application of International Conventions.
4. Excluded spaces are those which are open to the sea and, therefore, not suitable for the carriage of perishable cargoes.
5. Cargo spaces are defined as compartments for the transport of cargo which is to be discharged from the ship. They are to be permanently marked with the letters CC.
6. Alterations to the parameters of the net tonnage formula, which could result in a reduction of net tonnage, are restricted to once a year.

The application of this convention and the use of the above formulae has meant that some existing vessels, which had large exempted spaces, now have larger gross tonnages. Roll-on roll-off ships and ferries have had their gross and net tonnages significantly increased. Bulk carriers, ore carriers and other ships designed to carry high density cargoes, have had their net tonnage values reduced.

3.10 Other tonnage systems

A ship will carry a tonnage certificate indicating the values of tonnage for the vessel according to the International Convention. Other special tonnages exist, which are calculated in slightly different ways and are shown on special certificates. These are used for ships passing through the Suez and Panama canals. The charges levied for the use of these canals are based upon their particular canal tonnage measurement.

4 Calculation of areas and volumes

The three-dimensional shape of a ship's hull is not readily represented mathematically and calculations of areas and volumes are usually done using approximate rules. These same rules can be further applied to obtain centroids of area and volumes, moments and second moments of area (moments of inertia).

4.1 Trapezoidal rule

This rule approximates the area beneath a curve into a series of trapezoids, whose straight line boundaries enable the area within to be readily calculated. If part of the half waterplane area of a ship is considered, see Fig 4.1, its area can be approximated. If the area is split into two equal halves of length, h, and the ordinates are numbered y_0, y_1 and y_2, then:

$$\text{Area} = \frac{h(y_0 + y_1)}{2} + \frac{h(y_1 + y_2)}{2} = \frac{h(y_0 + 2y_1 + y_2)}{2}$$

The method can be extended to the complete half waterplane of a ship by selecting a suitable ordinate spacing to divide the ship's length into, say 10 equal parts, see Fig 4.2, and using the 11 ordinates to approximate the area:

$$\text{Area} = \frac{h(y_0 + 2y_1 + 2y_2 + 2y_3 + 2y_4 + 2y_5 + 2y_6 + 2y_7 + 2y_8 + 2y_9 + y_{10})}{2}$$

Where greater accuracy is required, eg, at the ends of a ship, additional ordinates can be used at half or quarter divisions.

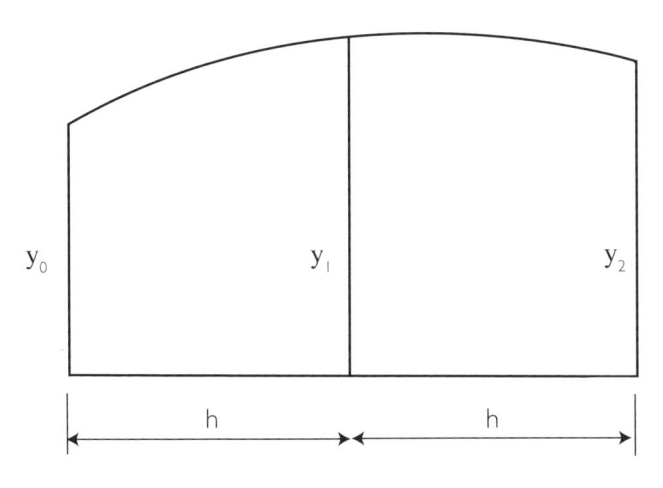

Fig 4.1: Trapezoidal rule

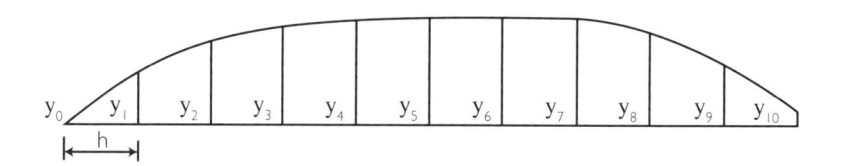

Fig 4.2: Trapeziodal rule applied to a half waterplane area

4.2 Simpson's rule

Simpson's rule replaces the curve enclosing an area by a mathematical curve of higher order. If three equally spaced ordinates, y_0, y_1 and y_2, are used for an area bounded by a curve and the distance between ordinates is h, then the area can be determined as follows:

$$\text{Area} = \frac{h}{3}(y_0 + 4y_1 + y_2)$$

This formula correctly calculates the area beneath a curve of second order and is widely used in calculations of area and volume.

If the area of a larger shape, such as a ship's half waterplane were required and the values for an odd number of ordinates were known, then the rule could be applied in a compound form, see Fig 4.3.

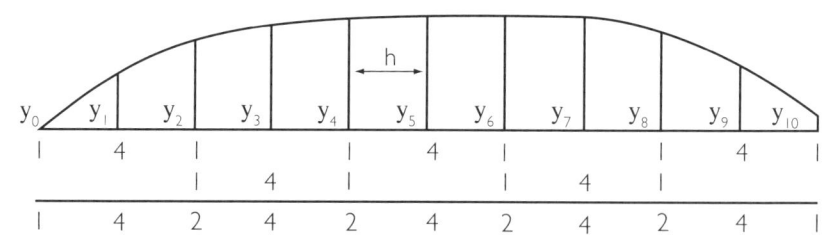

Fig 4.3: Simpson's rule applied to a half waterplane area

If the spacing between the ordinates is h, and the area between three ordinates is taken progressively along the curve, then:

$$\text{Total area} = \frac{h}{3}(y_0 + 4y_1 + 2y_2 + 4y_3 + 2y_4 + 4y_5 + 2y_6 + 4y_7 + 2y_8 + 4y_9 + y_{10})$$

about the ship's centreline, the ordinates are actually half-ordinates of the full waterplane and the area of the complete waterplane is therefore double the total area determined above.

Example

A ship 180m long has half-ordinates of 1, 7.5, 12, 13, 14, 14, 14, 13, 12, 7 and 0m. Calculate the waterplane area.

Half-ordinate	Simpson's multiplier	Product
1	1	1
7.5	4	30
12	2	24
13	4	52
14	2	28
14	4	56
14	2	28
13	4	52
12	2	24
7	4	28
0	1	0
	Product total, Σ_A	**323**

Common interval, $h = \dfrac{180}{10} = 18\text{m}$

Waterplane area $= \dfrac{h}{3} \times$ product total, $\sum_A, \times 2 = \dfrac{18}{3} \times 323 \times 2 = 3876\text{m}^2$

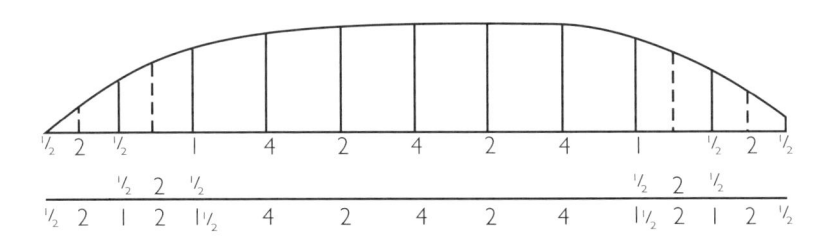

Fig 4.4: Use of half-ordinates at ends of a ship

Since there is a considerable change in area at the ends of a ship, then half-ordinates are usually used at $\frac{1}{2}$, $1\frac{1}{2}$, $8\frac{1}{2}$ and $9\frac{1}{2}$, see Fig 4.4. The mid-ordinate multipliers are then divided by the appropriate fraction to enable a continuous calculation:

Half waterplane area $= \dfrac{h}{3} (\frac{1}{2}y_0 + 2y_{1/2} + y_1 + 2y_{1\frac{1}{2}} + 1\frac{1}{2}y_2 + 4y_3 + 2y_4 + 4y_5$
$+ 2y_6 + 4y_7 + 1\frac{1}{2}y_8 + 2y_{8\frac{1}{2}} + y_9 + 2y_{9\frac{1}{2}} + \frac{1}{2}y_{10})$

Example
The half-ordinates of a cross-section through a ship are:

Waterline	Keel	0.25	0.50	0.75	1.0	1.5	2.0	2.5	3.0	4.0	5.0	6.0	7.0
Half-ordinate	2.9	5.0	5.7	6.2	6.6	6.9	7.2	7.4	7.6	7.8	8.1	8.4	8.7

Calculate the area of the cross-section to the 7m waterline.

Waterline	Half-ordinate	Simpson's multiplier	Product
Keel	2.9	$1/4$	0.73
0.25	5.1	1	5.10
0.50	5.7	$1/2$	2.85
0.75	6.2	1	6.20
1.00	6.6	$3/4$	4.95
1.5	6.9	2	13.80
2.0	7.2	1	7.2
2.5	7.4	2	14.8
3.0	7.6	$1 1/2$	11.4
4.0	7.9	4	31.6
5.0	8.1	2	16.20
6.0	8.4	4	33.60
7.0	8.7	1	8.70
		Product total, Σ_A	157.13

Area of cross-section

$$= \frac{h}{3} \times \text{Product total, } \sum\nolimits_A, \times 2 = \frac{1}{3} \times 157.13 \times 2 = 104.75 \text{m}^2$$

4.3 Application to volumes

So far, the ordinates put through Simpson's rule have represented length, and an area has been found for the shape. If the ordinates were to represent areas, for instance the waterplane areas of a ship, and these were put through Simpson's rule, with a common interval of draught, the volume of displacement of a ship could be found. Hold and tank capacities can be found in a similar manner.

4.4 Application to First and Second Moments of Area

Simpson's rule can be used to find the position of centroids of curved planes or Second Moments of Area of waterplanes. Consider the plane shown in Fig 4.5. If this is divided into thin strips of width dx and one particular strip at distance x from AB with an ordinate height, y, is considered, then:

Area of strip $= y dx$
Area of plane $= (y_1 + y_2 + y_3 + ...) dx$
$\quad\quad\quad\quad\quad = \Sigma y dx$

This area can be determined using Simpson's rule.

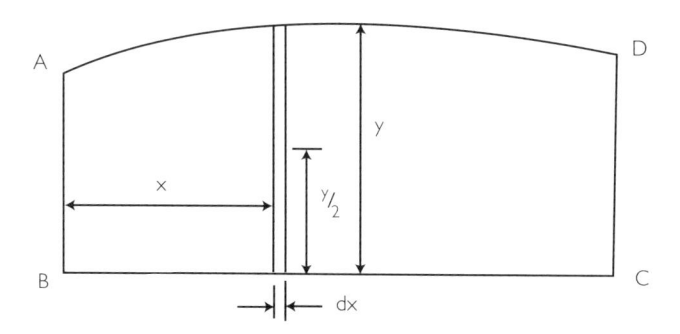

Fig 4.5: Simpson's rule applied to First and Second Moments of Area

If a First Moment of Area about axis AB is now considered, again with reference to Fig 4.5:

First Moment of Area of strip about AB = x × ydx = xydx
First Moment of Area of plane about AB = $(x_1y_1 + x_2y_2 + x_3y_3 + ...)$
$= \Sigma xydx$

If xy is considered as an ordinate value, then Simpson's rule may be used to determine the First Moment of Area of the plane.

If a Second Moment of Area about axis AB is now considered, again with reference to Fig 4.5:

Second Moment of Area of strip about AB = $I_{na} + Ax^2$ (using the parallel axis theorem)

$$= \frac{1}{12} y(dx)^3 + x^2 ydx$$

Since dx is small, then $(dx)^3$ can be ignored.

Second Moment of Area of plane about AB = $(x_1^2 y_1 + x_2^2 y_2 + x_3^2 y_3 + ...)dx$
$= \Sigma x^2 ydx = I_{yy}$

If x^2y is put through Simpson's rule then the Second Moment of Area can be found.

If the axis BC is now considered:
First Moment of Area of strip about BC = $\frac{1}{2}y \times ydx = \frac{1}{2}y^2 dx$
First Moment of Area of plane about BC = $\frac{1}{2}(y_1^2 + y_2^2 + y_3^2 + ...)dx$
$= \Sigma \frac{1}{2} y^2 dx$

If $\frac{1}{2}y^2 dx$ is put through Simpson's rule, then the First Moment of Area can be found.

Second Moment of Area of strip about $BC = \frac{1}{12} y^3 dx + (\frac{1}{2}y)^2 ydx = \frac{1}{3} y^3 dx$

Second Moment of Area of plane about BC = $\frac{1}{3}(y_1^3 + y_2^3 + y_3^3 + ...)dx$
$= \Sigma \frac{1}{3} y^3 dx = I_{xx}$

Again this value may be found by putting $\frac{1}{3}y^3 dx$ through Simpson's rule.

4.5 Summary

Area of plane $= \Sigma y dx$
First Moment of Area about AB $= \Sigma xy dx$
First Moment of Area about BC $= \Sigma \frac{1}{2}y^2 dx$
Second Moment of Area about AB $= \Sigma x^2 y dx = I_{YY}$
Second Moment of Area about BC $= \Sigma \frac{1}{3}y^3 dx = I_{XX}$

First Moments of Area and centroids

The calculation of a First Moment of Area is usually done to determine the position of the centroid of the area:

Distance of centroid (centre) from a given axis

$$= \frac{\text{First Moment of Area about the axis}}{\text{Area}}$$

Moments of area are usually required about two principal axes, the longitudinal and the transverse.

The longitudinal position of a centroid of a waterplane is usually found in relation to midships. In this way, lever distances measured from midships may be used and the moments forward and aft found. The position of the centroid forward or aft is determined by the net moment value.

The longitudinal centroid position is known as the **Centre of Flotation**, F, or the **Longitudinal Centre of Flotation,** LCF.

Example:

The half-ordinates of a waterplane 180m long are:

Section	AP	1	2	3	4	5	6	7	8	9	FP
Half-ordinate	0	8.0	10.5	12.5	13.5	13.5	12.5	11	7.5	3.0	0

Calculate:
1. Area of waterplane.
2. Position of LCF from midships.
3. Second Moment of Area of waterplane about a transverse axis through the LCF.
4. Second Moment of Area about the centreline.

Calculations for transverse axes

Section	Half-ordinate	Simpson's multiplier	Product for area	Lever	Product for first moment	Lever	Product for second moment
0	0	1	0	5	0	5	0
1	8.5	4	34	4	136	4	544
2	10.5	2	21	3	63	3	189
3	12.5	4	50	2	100	2	200
4	13.5	2	27	1	27	1	27
5	13.5	4	54	Total	326	-	0
6	12.5	2	25	-1	-25	-1	25
7	11	4	44	-2	-88	-2	176
8	7.5	2	15	-3	-45	-3	135
9	3.0	4	12	-4	-48	-4	192
10	0	1	0	-5	0	-5	0
		Total	282	Total	-206	Total	1488

Calculations for centreline axis

Section	Half-ordinate	(Half-ordinate)³	Simpson's multiplier	Product for second moment
0	0	0	1	0
1	8.5	614.1	4	2456.4
2	10.5	1157.6	2	2315.2
3	12.5	1953.1	4	7812.4
4	13.5	2460.4	2	4920.8
5	13.5	2460.4	4	9841.6
6	12.5	1953.1	2	3906.2
7	11	1331.0	4	5324.0
8	7.5	421.9	2	843.8
9	3.0	27	4	108
10	0	0	1	0
			Total	37 528.4

Area of waterplane = $^2/_3 \times 18 \times 282 = 3384 m^2$

LCF from midships $= \dfrac{(326-206)}{282} \times 18 = 7.66 m$ aft

Second Moment of Area about midships = $^2/_3 \times 18^3 \times 1488 = 5\ 785\ 343 m^4$

Second Moment of Area about LCF = $5\ 785\ 343 - (3384 \times 7.66^2) = 5\ 586\ 785 m^4$

Second Moment of Area about the centreline = $^2/_9 \times 18 \times 37\ 528.4 = 150\ 113 m^4$

5 Buoyancy, stability and trim

The two principal forces which act on a ship floating freely are weight and buoyancy. For the ship to float, it must displace its own weight of water, and for equilibrium, the centres of weight and buoyancy must be vertically in line. External or internal forces can move the ship in either a transverse or longitudinal direction, and its ability to return to the equilibrium position is related to stability. The ability of a ship to remain afloat after damage is also related to buoyancy.

5.1 Buoyancy and displacement

When a ship of mass, m, is floating freely in water there is a force acting downwards, due to gravity, equal to mg, where g is the acceleration due to gravity. Since the ship floats and does not sink, there must be an equal and opposite force acting upwards, known as buoyancy. Archimedes' principle states that when a solid is immersed in a liquid, it experiences an upthrust equal to the weight of the fluid displaced. For a body to float, the weight of the body must equal the weight of the fluid displaced. The mass of a ship can be found by calculating the underwater volume and multiplying by the density of the water in which it is floating. Mass, rather than weight, is used in most naval architecture calculations, and the mass of a ship is often referred to as the displacement. The centre of mass is generally known as the centre of gravity, G, and the centroid of the underwater volume is known as the centre of buoyancy, B. For a floating ship to be in equilibrium, B and G must lie in the same vertical line in both the longitudinal and transverse axes.

The underwater volume of a ship can be calculated by applying Simpson's rule to the immersed areas of a number of sections along the length. Where the immersed cross-sectional areas are calculated to a number of waterlines drawn parallel to the base of the ship, then underwater volumes at each of these waterlines can be found.

Example
A ship floats at an even keel draught of 6.25m and the areas of waterplanes at 1.25m intervals are:

Waterplane m above baseline	1.25	2.5	3.75	5	6.25
Area m^2	1450	1650	1730	1780	1800

The hull, below the 1.25m waterline, has a displacement of 1500t and a vertical centre of buoyancy, KB, of 0.75m.

Find the displacement of the ship, in seawater of density $1.025 kg/m^3$, and the position of the vertical centre of buoyancy of the ship.

Waterplane m	Area m²	Simpson's multiplier	Product for area	Lever above base	Moment for product of area
1.25	1450	1	1450	1	1450
2.5	1650	4	6600	2	13 200
3.75	1730	2	3460	3	10 380
5	1780	4	7120	4	28 480
6.25	1800	1	1800	5	9000
		Total	20 430	Total	62 510

$$\text{Displacement of hull} = \frac{1}{3} \times 1.25 \times 20\,430 \times 1.025$$
$$= 8725.3 \text{ tonnes}$$

$$\text{KB of hull} = \frac{62\,510}{20\,430} \times 1.25 = 3.83\text{m}$$

Item	Displacement	KB	Moment
Hull (from 1.25 to 6.25m)	8725.3	3.83	33 417.9
Lower hull (0 to 1.25m)	1500	0.75	1125
Total	10 225.3	Total	34 542.9

$$\text{Centre of buoyancy, KB, of ship} = \frac{34\,542.9}{10\,225.3} = 3.38\text{m}$$

5.2 Bonjean curves

Underwater volume can also be found by reference to Bonjean curves. Curves of immersed cross-sectional area are plotted against draught and usually superimposed on the profile of the ship as shown in Fig 5.1. If a particular waterline, W_1L_1, is drawn on the profile, the area at the section is found by extending the intercept to the Bonjean curve for that section. Bonjean curves are particularly useful for waterlines which are not parallel to the base.

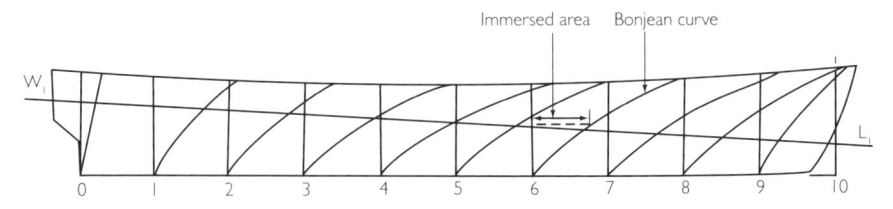

Fig 5.1: Bonjean curves

5.3 Transverse stability

With reference to ships, stability is the ability to return to the original position after being inclined. As mentioned earlier, a ship floats in equilibrium under the action of two forces, the weight acting downwards and the buoyancy acting upwards. Also the centre of gravity, G, must be vertically in line with the centre of buoyancy, B.

When a ship is inclined, from initial waterline, WL, to some new waterline, W_1L_1, by an external force, such as the wind or waves, the centre of gravity, G, will remain in the same position. The underwater shape of the hull will, however, change and the centre of buoyancy, B, will move to a new position, B_1, see Fig 5.2. This creates a moment W × GZ, which tends to return the ship to the upright position. The product W × GZ is known as the **righting moment** and GZ is known as the **righting lever**. Since this moment tends to right the ship, it is then said to be stable. For small angles of inclination, θ, or heel, up to about 5deg, the vertical through B_1 – the new centre of buoyancy – intersects the centreline at M, the transverse metacentre.

Thus   GZ = GMsinθ

It can be seen that GZ is a function of GM, but GM is independent of θ. The initial stability of a ship is usually expressed in terms of GM, the **metacentric height**. A ship with a small metacentric height will have a small righting lever and will thus roll easily; such a ship is said to be **tender**. A ship with large metacentric height will have a large righting lever and therefore a considerable resistance to rolling; such as ship is said to be **stiff**. A stiff ship will have an uncomfortable rolling motion and structural damage may occur as result of the small rolling period.

If the centre of gravity were to lie above the transverse metacentre, then the moment would act to increase the angle of heel. A ship in this condition would be unstable and will not return to the upright position after being inclined. In this case the metacentric height is considered to be a negative value. When G and M lie at the same point there will be no righting lever and the ship would simply remain at the angle to which it was inclined. This condition is known as neutral equilibrium. It is not a practical situation since the slightest change in the position of G will create stability or instability.

5.4   Determining the transverse metacentre

The height of the transverse metacentre, KM, is given by KM = KB + BM. It was shown earlier how to determine KB. A method of determining BM is therefore required.

A ship of volume of displacement, ∇, is initially upright at waterline WL, with the centre of buoyancy B on the centreline, see Figure 5.3. If the ship is now inclined through a small angle, θ, to some new waterline, W_1L_1, which intersects WL at S on the centreline, a wedge of buoyancy, WSW_1, has moved across to L_1SL and the centre of buoyancy will move from B to B_1.

Let v = volume of wedge,
    gg_1 = transverse shift of centre of gravity of wedge,

then $BB_1 = \dfrac{v \times gg_1}{\nabla}$

but $BB_1 = BM \tan \theta$

thus $BM \tan \theta = \dfrac{v \times gg_1}{\nabla}$

$$BM = \dfrac{v \times gg_1}{\nabla \tan \theta}$$

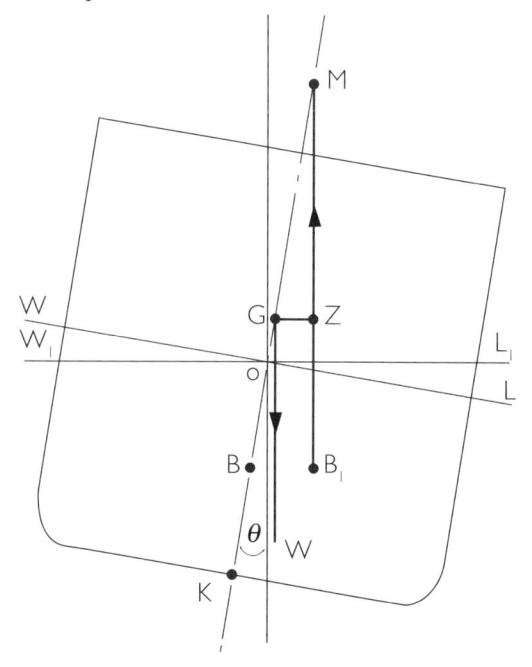

Fig 5.2: Small angle stability

To determine the value of v × gg₁, the ship's length is divided into thin strips of length dx. The half width of a strip will be y.

Volume of strip of wedge = $\frac{y}{2} \times y \tan\theta \, dx$

Moment of shift of strip of wedge $= \frac{4y}{3} \times \frac{y^2 \tan\theta}{2}$

$$= \frac{2y^3 \tan\theta \, dx}{3}$$

Total moment of shift of wedge = v × gg₁

$$= \sum \frac{2y^3}{3} \tan\theta \, dx$$

$$= \tan\theta \frac{2}{3} \sum y^3 dx$$

but $\frac{2}{3} \sum y^3 dx$ = second moment of area of a waterplane about the centreline of the ship = I_{xx}

Thus $v \times gg_1 = I_{xx} \tan\theta$

$$BM = \frac{I_{xx} \tan\theta}{\nabla \tan\theta}$$

hence $BM = \frac{I_{xx}}{\nabla}$

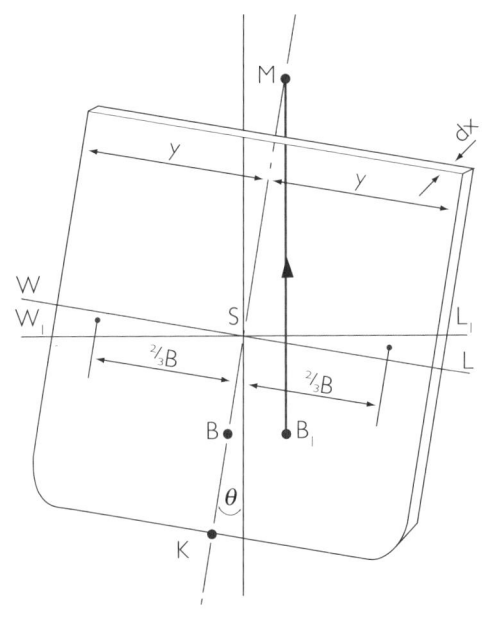

Fig 5.3: Determining the transverse metacentre

The height of the centre of buoyancy above the keel, KB, is the height of the centroid of the underwater volume above the base and, by adding BM, the height of the transverse metacentre, KM, can be found. Where the height of the centre of gravity above the base, KG, is known, then the metacentric height, GM, can be found.

The values of BM, KB and, hence, KM can be readily found for simple geometric forms, such as a rectangle or a triangular cross-section.

Example
A vessel of constant rectangular cross-section has a breadth of 20m and a metacentric height of one-third of the draught. The vertical centre of gravity lies on the waterline. Calculate the draught.

Let T = draught
Then KB = T/2; KG = T and GM = T/3

Now $BM = \dfrac{I}{\nabla} = \dfrac{1}{12} \times \dfrac{L \times B^3}{L \times B \times T} = \dfrac{B^2}{12T}$

Since KG + GM = KB + BM

Then $T + \dfrac{T}{3} = \dfrac{T}{2} + \dfrac{20^2}{12T}$

Draught, T = 6.33m

5.5 Metacentric diagram

The height of the centre of buoyancy above the keel, KB, and the height of the transverse metacentre, KM, depend upon draught and the values for a ship for a range of draughts can be calculated and plotted to form a metacentric diagram, see Fig 5.4.

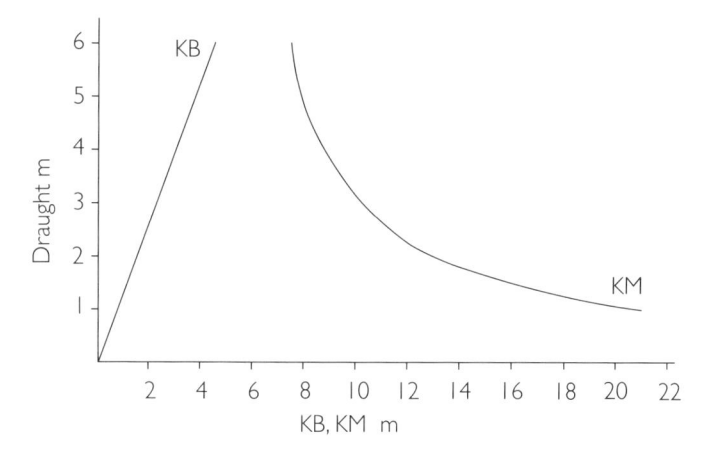

Fig 5.4: Metacentric diagram

5.6 Determining the centre of gravity

An inclining experiment is carried out on new ships, and those that have had large structural alterations made, in order to initially find the centre of gravity, KG, and ultimately the metacentric height, GM, and thus establish that the ship is stable.

During the experiment, one or more masses are moved through a known distance, transversely across the deck. The resulting angle of heel, θ, is measured by means of two pendulums, one at each end of the ship, or an instrument known as a stabilograph. Where a pendulum is used, the deflection is measured against a batten. Pendulums should be as long as possible, to improve accuracy of measurement, and of fine wire with a heavy bob to ensure free movement. The bob may be immersed in oil or water, to dampen its movement.

As the mass, w, is moved across the deck through a distance, l, the centre of gravity of the ship moves from G to G_1, see Fig 5.5.
$w \times l = W \times GG_1$ where W is the displacement of the ship.

The ship will incline to some angle, θ, and will be in stable equilibrium. The centre of buoyancy, originally at B, will move to B_1, vertically beneath G_1. The movement of GG_1 is parallel to the movement of w, therefore the angle G_1GM is a right angle.

$GG_1 = GM \tan\theta$

$$\frac{w \times l}{W} = GM \tan\theta$$

thus $GM = \dfrac{w \times l}{W \tan\theta}$

now $\tan\theta = \dfrac{\text{deflection of pendulum}}{\text{length of pendulum}} = \dfrac{GG_1}{GM}$

The only unknown value is the metacentric height, GM, which can then be found.

5.7 Conduct of the inclining experiment

An inclining experiment is used to determine the position of a ship's centre of gravity, since to determine it from the individual masses of steel and equipment would be a difficult, if not impossible, calculation. The position of a ship's centre of gravity is critical in assessing its stability and thus the accuracy of the inclining experiment is very important. The inclining experiment is conducted on the ship when it is as near complete as possible and in the almost lightweight condition. Adjustments can then be made to establish the centre of gravity in the lightweight condition and subsequently for any loaded condition of the ship, when the masses and centres of gravity of cargoes which have been loaded and unloaded are known.

Two sets of weights are generally used, one on each side of the ship, and the inclination for the movement of each weight both to port, then back again and then to starboard and back again is noted. The draughts of the ship are measured and so is the specific gravity of the water in which it is floating, in order to determine the displacement.

The metacentric height, GM, can be found as outlined earlier. The height of the metacentre from the keel, KB, can be found from hydrostatic curves for the ship at the mean draught and thus the height of the centre of gravity KG can be found.

This value will have to be corrected for any items which are not yet fitted to the ship and those which make up the lightweight, eg water in boilers.

Certain precautions are necessary in the conduct of the experiment to ensure as accurate a result as possible:

1. A calm day with no wind should be chosen.
2. The ship should be free to move with all moorings slack.
3. Any loose items should be secured.

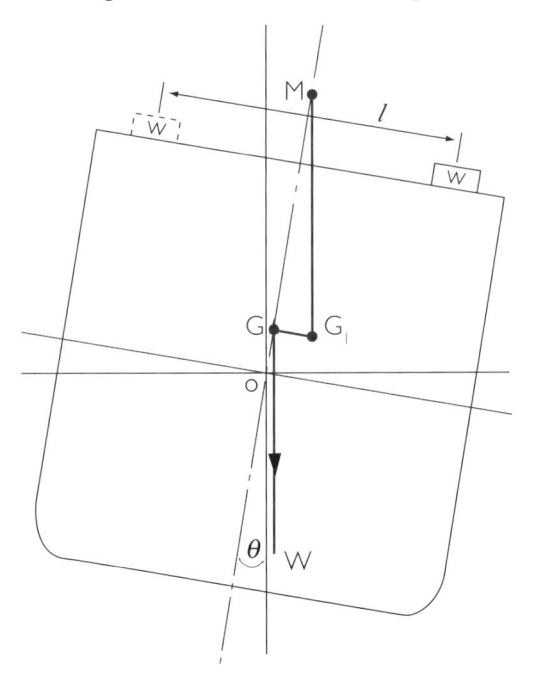

Fig 5.5: Determining the centre of gravity

4. All tanks should be empty or full to avoid free surface effect.
5. As few personnel as possible should be on board and they should be on the centreline when readings are taken.
6. A large trim should be avoided.

Example
A ship of 9000t displacement and height of metacentre, KM, of 7m, was inclined by moving 4t through 18m across the deck. The deflection of a 10m pendulum was seen to be 125mm.

The following items were on board at the time of the experiment and do not form part of the lightweight.

Item	Mass (tonnes)	Vertical centre of gravity m
Inclining masses	16	12.2
Oil fuel	100	9.4
Fresh water	70	10.7
Water ballast	180	6.1
Miscellaneous	40	11.6

Calculate the lightweight of the ship and its vertical centre of gravity, KG. (Note that since all masses are in tonnes, the moment will be tonne-m. All masses could be converted into the SI unit of weight, MN, and the moments would then be in MNm)

$$\text{GM as inclined} = \frac{\text{inclining mass} \times \text{distance moved}}{\text{displacement} \times \tan\theta}$$

$$= \frac{4 \times 18}{9000 \times 125 / 10\,000} = 0.64\,\text{m}$$

Centre of gravity as inclined = 7 - 0.64 = 6.36m

Mass (tonnes)	Centre of gravity, KG m	Moment tonne-m
9000	6.36	57 240
16	12.2	195.2
100	9.4	940
70	10.7	749
180	6.1	1098
40	11.6	464
8594(Lightweight)		53 793.8 (Lightweight moment)

Note: The various masses are deducted from 9000 to give the lightweight. Their respective moments are deducted from the vessel's moment to give the lightweight moment.

Lightweight = 8594 tonnes

Centre of gravity for lightweight $= \dfrac{53793.8}{8594} = 6.26\,\text{m}$

5.8 Operations affecting stability

A number of ship operations can affect stability and their effects must be understood and, where possible, mitigated. When liquid is consumed or removed from tanks then a 'free surface' is created which can move and affect stability. When a weight is lifted from a deck and suspended, its centre of gravity rises to the point of suspension, with a subsequent effect upon stability. When a single piece of solid cargo, or a quantity of loose dry bulk cargo, moves transversely across the ship, it will list to one side with some loss of stability. Wind can also influence stability, and the effect would be similar to the transverse movement of a weight in the direction in which the wind is blowing. When a ship descends onto the supporting blocks of a drydock and thus moves from a floating, to a ground supported situation, the conditions for stability change. This operation will be considered in a later section, since it involves both transverse stability and trim.

5.9 Free surface correction

A tank on the ship's centreline and partly full of liquid will affect the stability of a ship when it heels or rolls. The body of liquid moves with the ship's movement and the centre of gravity of the liquid will move away from midships in the direction of the roll. The ship's centre of gravity will move and thus reduce the righting lever, GZ. This is known as free surface effect.

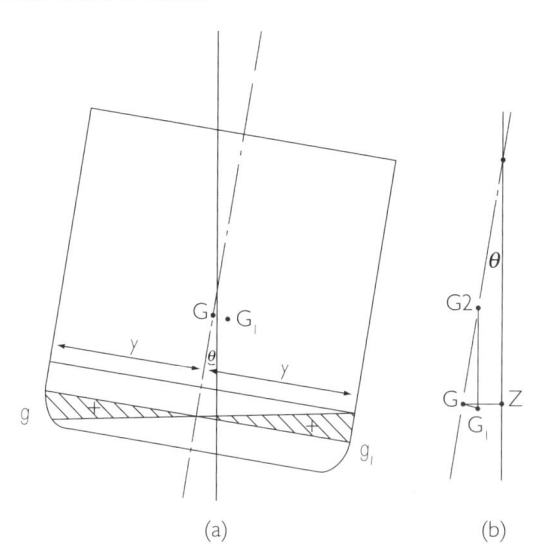

Fig 5.6: (a) and(b): Free surface effect

Consider a tank partly full of water in a ship of displacement, W, inclined to some angle θ, as shown in Fig 5.6(a). The ship's centre of gravity will move from G to G_1 as the wedge of liquid moves across the tank.

$$\text{Thus } GG_1 = \frac{m \times gg_1}{W}$$

where m = mass of wedge = volume of wedge, v × density of liquid, ρ_l
W = displacement of ship = volume of displacement, ∇ × density seawater, ρ_{sw}

$$GG_1 = \frac{v \times \rho_l \times gg_1}{\nabla \times \rho_{sw}}$$

If the tank is divided into strips of length dx and the wedge of liquid is assumed to remain triangular:

Volume of strip of wedge = $\frac{1}{2}y \times y\tan\theta dx = \frac{1}{2}y^2\tan\theta dx$

Mass of strip of wedge = $\rho_l \times \frac{1}{2}y^2\tan\theta dx$

Moment of transfer of strip = $\frac{4}{3}y \times \rho_l \times \frac{1}{2}y^2\tan\theta dx$
$= \rho_l \times \frac{2}{3}y^3\tan\theta dx$

Total moment of transfer of wedge = $v \times \rho_l \times gg_1 = \rho_l\tan\theta\Sigma\frac{2}{3}y^3\tan\theta dx$

But $\Sigma\frac{2}{3}y^3\tan\theta dx$ = Second Moment of Area of free surface about the centreline of the tank = I_{xx}

Thus $GG_1 = \dfrac{\rho_l I_{xx} \tan\theta}{\rho_{sw}\nabla}$

As a result of this movement of the ship's centre of gravity, the righting lever GZ is reduced to G_1Z, which is equal to G_2Z, see Fig 5.6 (b). The effect of the loose liquid in the tank is equivalent to a reduction in metacentric height, GM, of GG_2 where

$$GG_1 = GG_2 \tan\theta = \frac{\rho_l I_{xx} \tan\theta}{\rho_{sw} \times \nabla \times \tan\theta} = \frac{\rho_l \times I_{xx}}{\rho_{sw} \times \nabla}$$

5.10 Use of tank divisions

Free surface effect can be minimised by the use of longitudinal tank divisions; these effectively reduce the value of the second moment of area of the free surface about the centreline of the tank. If a single centreline division is used, the free surface effect or reduction in metacentric height is reduced to one-quarter. If two longitudinal divisions are used to create three equal-width tanks, the free surface effect is reduced to one ninth.

Example
A ship of 10 000t displacement has a rectangular double bottom tank 8m long and 12m wide. Calculate the free surface effect if this tank is partly full of liquid of density 0.85t/m³. Determine the reduction in free surface effect if a longitudinal centreline division is fitted in the tank.

Second moment of free surface about the centreline of the tank = $\frac{1}{12} \times 8 \times 12^3$ = 1152m⁴

$$GG_2 = \frac{\text{density liquid} \times I_{xx}}{\rho_{sw}\nabla} = \frac{0.85 \times 1152}{10\,000} = 0.09792\text{m}$$

If a longitudinal centreline division is fitted:
for each new tank of 6m wide, I_{xx} of free surface about centreline of the tank = $\frac{1}{12} \times 8 \times 6^3 = 144$m⁴
thus, total I_{xx} for both tanks = 288m⁴

$$GG_2 = \frac{\text{density liquid} \times I_{xx}}{\rho_{SW} \nabla} = \frac{0.85 \times 288}{10\,000} = 0.02448 \text{m}$$

Reduction in free surface effect = 0.09792 − 0.02448 = 0.07344m

5.11 Effect of suspended weights on stability

If a weight, w, is suspended from a point distance, h, from its centre of gravity and is free to move, it will affect a ship's stability. When the ship heels to some angle, θ, the weight will move transversely to a new position with its centre of gravity at the point of suspension.

Transverse movement of weight = w × hsinθ
Transverse movement of ship's centre of gravity, $GG_1 = \dfrac{wh\sin\theta}{W}$

where W is the displacement of the ship.
Loss of righting moment = W × GG_1 = wh sinθ

This loss can be viewed in the same way as that due to free surface effect, where G is raised to some new position G_2, vertically above G,
thus

$$GG_2 = \frac{GG_1}{\tan\theta} = \frac{wh\sin\theta / W}{\tan\theta} = \frac{wh}{W}$$

since sinθ = tanθ for small angles.

The movement of cargo by a shipboard crane is an example of this effect, which must be borne in mind by the ship operator. It is particularly important to consider this effect when lifting very heavy items of cargo.

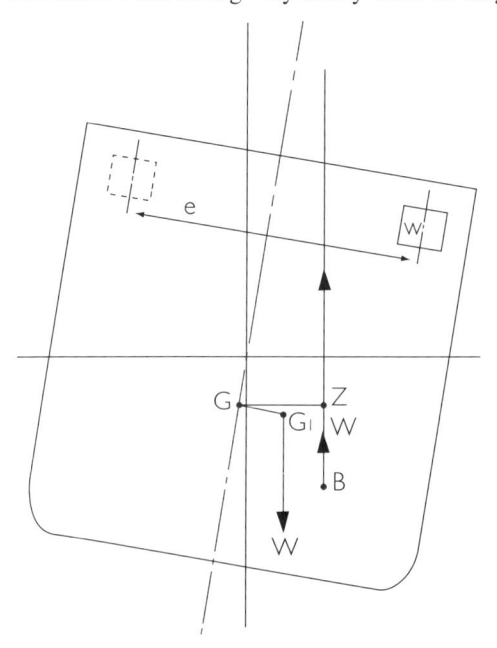

5.12 Transverse movement of weight

The permanent movement of a single item of cargo, possibly due to it breaking loose, or the shifting of bulk cargo, can reduce a ship's centre of gravity. Consider an item of cargo of weight, w, which moves a transverse distance, l. The ship's centre of gravity will move from G to G_1, see Fig 5.7.

Transverse movement of centre of gravity, $GG_1 = \dfrac{w \times l}{W}$

where W is the weight of the ship.

The righting lever, GZ, will be reduced by $GG_1\cos\theta$

Fig 5.7: Transverse movement of weight

Reduced righting lever = $GZ - \dfrac{wl\cos\theta}{W}$

This reduction will occur on the side to which the weight moved, but there will be an increase on the opposite side of the ship. The $GG_1\cos\theta$ value can be plotted on a curve of statical stability and where it intersects will be the angle of list at the equilibrium point (nearest to the origin) and the range of stability (now reduced). The maximum value of GZ will also be reduced.

Dry bulk cargoes are a special case of this situation since, although solid, they can have fluid-like movements. A loaded cargo will settle and spaces will exist at the sides of the holds as the cargo takes up an **angle of repose**. This is the angle at which a pile of the cargo would settle in an open space. This cargo could move, and either special arrangements must be made to ensure the holds are full, or some form of longitudinal hold divisions or shaping of the upper hold is required.

5.13 Large angle stability

Stability has been considered so far in relation to small angles of inclination, up to about 5deg, where certain assumptions are made and the metacentric height, GM, is used as the measure of stability.

For larger angles of inclination the upright and inclined waterplanes cannot be assumed to intersect on the waterline, and the metacentre, M, does not remain in a fixed position on the centreline. The righting lever, GZ, which is the perpendicular distance between vertical lines through the centre of gravity and the inclined centre of buoyancy, is instead used as a measure of stability.

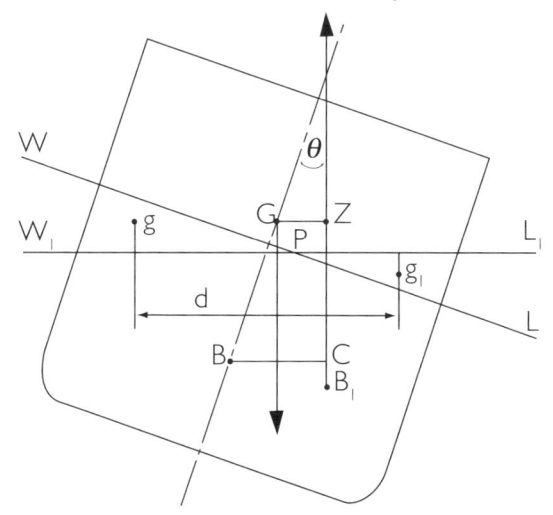

Fig 5.8: Large angle stability

Consider a ship which is inclined to some large angle θ from the vertical, see Fig 5.8. If WL is the initial waterline and W_1L_1 the waterline when inclined, the volume of displacement in each case will be the same. If the sides of the ship were vertical along its length then the two waterlines would intersect at the centreline. However, the ship's side is often at an angle to the waterline, particularly in the region of the bow, and thus the waterlines will intersect at some point, P. The volume WPW_1, which has emerged will be equal to the volume L_1PL, which has been immersed. If this volume is considered to be v and the centroids of each are denoted by g and g_1 located a distance, d, apart on the waterline W_1L_1, the horizontal movement of the centre of buoyancy can be found.

Thus $BC = \dfrac{v \times d}{\nabla}$

where ∇ is the volume of displacement of the ship.

Also $GZ = BC - BG \sin \theta = \dfrac{v \times d}{\nabla} - BG \sin \theta$.

This equation is called **Atwood's formula**, after the work done by George Atwood in the eighteenth century. If v and d are evaluated for a range of angles of inclination, a curve of righting lever, GZ, to a base of angle of inclination, θ, can be drawn, as shown in Fig 5.9. This is called a **curve of statical stability**. The righting lever can be seen to rise to a maximum value and then fall to zero. A ship inclined beyond the point of zero GZ will be unstable. The angle up to the zero GZ point is the **range of stability** of the ship, at that particular loaded condition. Ship operators need to know the maximum value of GZ and the range of stability for a range of loaded conditions.

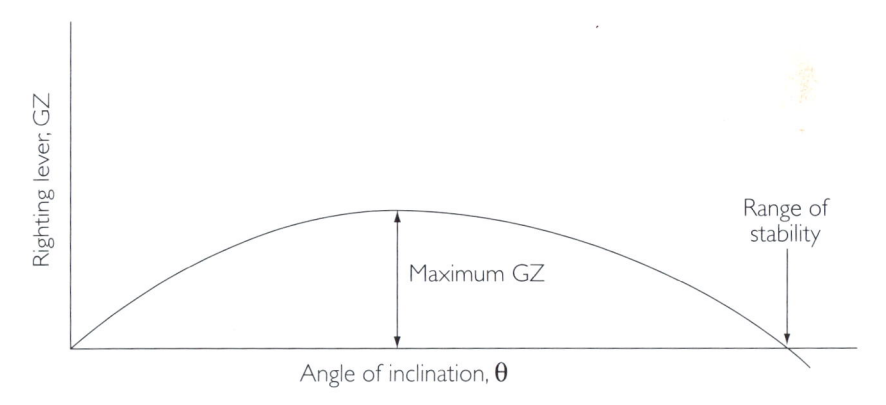

Fig 5.9: Curve of statical stability

5.14 Wall-sided ships

Where the sides of a ship are vertical or almost so, the righting lever, GZ, can be found accurately over a range of inclination up to about 15deg, or until the turn of bilge is exposed or the deck edge immersed.

Consider a ship initially at waterline WL, inclined to some angle θ and new waterline W_1L_1, see Fig 5.10. Since the ship is wall-sided the volume $W0W_1$ = volume $L0L_1$. The centre of buoyancy, B, will have moved to B_1, whose co-ordinates are α and β. The value of these co-ordinates can be calculated by examining the moment of transfer of the volume of the wedges in a horizontal and a vertical direction.

Transverse moment of volume $= \int_0^L \dfrac{1}{2} y \times y \tan \theta \, dx \times \dfrac{4}{3} y$

$= \int_0^L \dfrac{2}{3} y^3 \tan \theta \, dx = \tan \theta \int_0^L \dfrac{2}{3} y^3 \, dx$

$= I_{xx} \tan \theta$

where I_{xx} is the second moment of area of the waterplane about the centreline. Thus $\alpha = I_{xx} \tan\theta / \nabla = BM \tan\theta$, since $BM = I_{xx}/\nabla$.

Vertical moment of volume $= \int_0^L \frac{1}{2} y^2 \tan\theta \times \frac{2}{3} y \tan\theta \, dx$

$= \frac{1}{2} I_{xx} \tan^2\theta$

thus $\beta = \frac{1}{2} I_{xx} \tan^2\theta / \nabla = \frac{1}{2} BM \tan^2\theta$

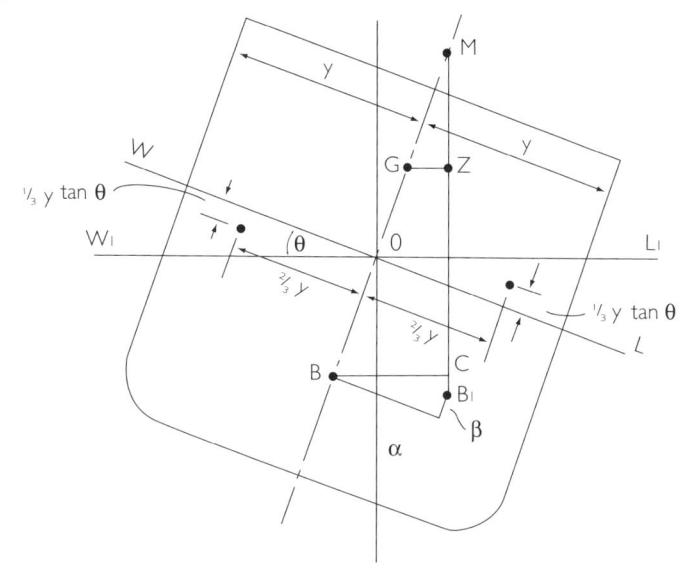

Fig 5.10: Wall-sided ships

By reference to Fig 5.10 it can be seen that:
$$BC = \alpha\cos\theta + \beta\sin\theta$$
$$= BM \tan\theta \cos\theta + \frac{1}{2} BM\tan^2\theta \sin\theta$$
$$= \sin\theta(BM + \frac{1}{2} BM\tan^2\theta)$$

It can also be seen that:
$$GZ = BC - BG\sin\theta$$
$$= \sin\theta (BM - BG + \frac{1}{2} BM \tan^2\theta)$$
$$= \sin\theta (GM + \frac{1}{2} BM \tan^2\theta)$$

This equation is often referred to as the **wall sided formula** and can be used for large angle stability calculations up to about 15deg, as long as the deck edge is not immersed or the round of bilge is not exposed. The difference between this expression for GZ and that for small angle stability is the addition of the term $\frac{1}{2} BM\tan^2\theta$.

This wall sided formula can be used to examine the stability of a ship. Consider first a ship with a positive metacentric height:
the ship will be in equilibrium when GZ = 0, thus

$0 = \sin\theta(GM + \frac{1}{2} BM\tan^2\theta)$

which has two solutions, either $\sin\theta = 0$, or $GM + \frac{1}{2} BM\tan^2\theta = 0$

The first equation is satisfied when θ = 0, ie the ship is upright. The second equation has no solution.

If the GM is zero:
the ship will be in equilibrium when GZ = 0, thus

$$0 = \tfrac{1}{2} BM\sin\theta\tan^2\theta$$

The only solution to this equation is θ = 0, ie the ship is upright. It should, however, be noted that if a ship with neutral stability is inclined at an angle there will be a righting lever, GZ, returning it to the vertical position, where

$$GZ = \tfrac{1}{2} BM\sin\theta\tan^2\theta$$

If the GM is negative:
the ship will be in equilibrium when GZ = 0
$$0 = \sin\theta\,(-GM + \tfrac{1}{2} BM\tan^2\theta)$$

This equation has two solutions:
$\sin\theta = 0$, ie θ = 0,
$\tan\theta = \pm\sqrt{2GM/BM}$.

The first solution results in instability, since any inclination will result in a lever continuing the movement away from the upright. The two solutions to the second equation give positions of stable equilibrium, since the ship will heel either to port or starboard to an angle, θ, where:

$$\theta = \pm \tan^{-1}\sqrt{2GM/BM}$$

This is known as the **angle of loll**. A ship with a negative metacentric height will not necessarily capsize when inclined; as implied by the theory of small angle stability, it will move to the angle of loll either to port or starboard. Waves acting against the ship could make it loll alternately to port and starboard, a clear indication of a lack of stability. A curve of righting lever, GZ, against angle of inclination, θ, plotted using the wall sided formula, would be as shown in Fig 5.11. Where the curve crosses the base line at the angle of loll, the ship would be stable, although inclined.

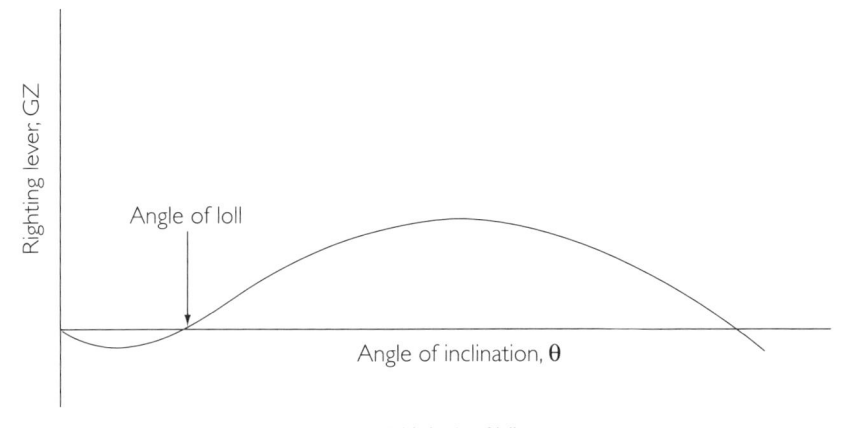

Fig 5.11: Angle of loll

A vessel with a negative metacentric height of 0.05m and BM of 6m would have an angle of loll of 7.37deg, a large angle for a small negative metacentric height. Furthermore, it can be shown that the metacentric height in the lolled position is effectively twice that in the upright position, but of opposite sign, unless the angle of loll is large.

The wall sided formula is a special case of Atwood's formula, which gives an expression for GZ when a ship is inclined at a large angle and assumes that the immersed and emerging volumes of water intersect on the centreline. The expression is difficult to evaluate, since on actual ships when inclined, the volumes of immersed and emerging volumes of water do not intersect on the centreline, but at some unknown point.

5.15 Cross curves of stability
Cross curves of stability are another way of determining the righting lever, GZ. They are curves of the righting lever GZ, drawn to a base of displacement for constant angles of heel. In effect, the ship is considered to be inclined to some angle θ, and values of GZ are found for various displacements. The centre of gravity, G, would vary with different displacements, but is assumed in a fixed condition for the calculations. A correction is ultimately necessary for the actual value of centre of gravity, G. Typical cross curves of stability are shown in Fig 5.12.

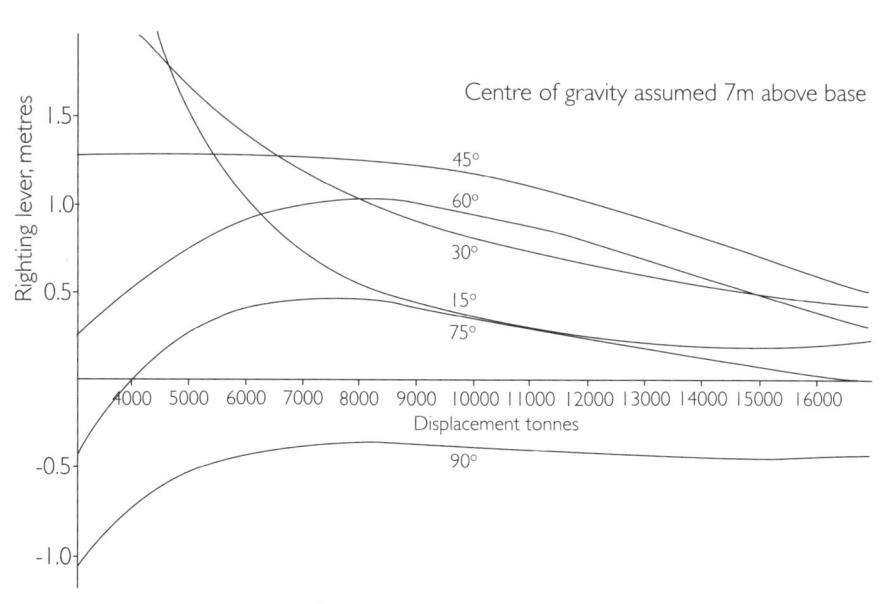

Fig 5.12: Cross curves of stability

5.16 Curves of statical stability
The cross curves of stability can be used to draw a curve of statical stability for a ship. This is the value of righting lever, GZ, at any angle of heel for a particular displacement and height of centre of gravity, KG. The curve is obtained by drawing a vertical line, at the required displacement value, on the cross curves and the ordinate to each curve is measured. The values obtained are then drawn on a base of angle of inclination and the curve of statical stability produced.

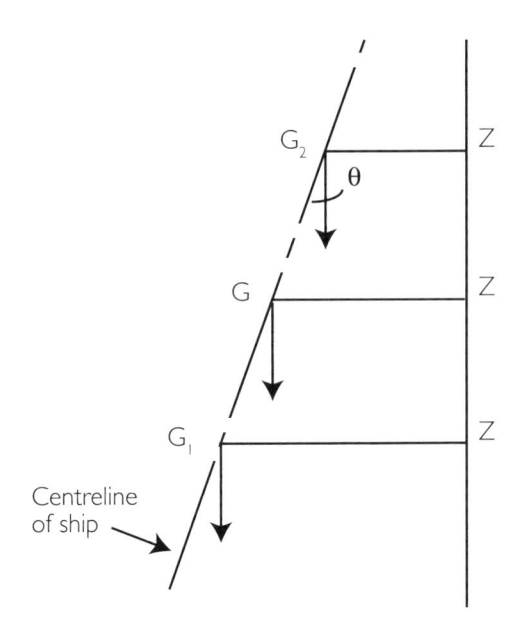

Fig 5.13: Amending cross

This curve can be amended for a different value of vertical height of centre of gravity, KG, to that used for the original cross curves, see Fig 5.13. The centre of gravity assumed for the cross curves is G. The actual positions for two different loading positions, at the same displacement and angle of inclination, are G_1 and G_2. It will be evident that G_1Z is larger than GZ and G_2Z is smaller than GZ.

Thus where the actual value, G_1, is below the assumed value of G, then
$G_1Z = GZ + GG_1\sin\theta$
and where the actual value, G2, is above the assumed value of G, then
$G_2Z = GZ - GG_2\sin\theta$

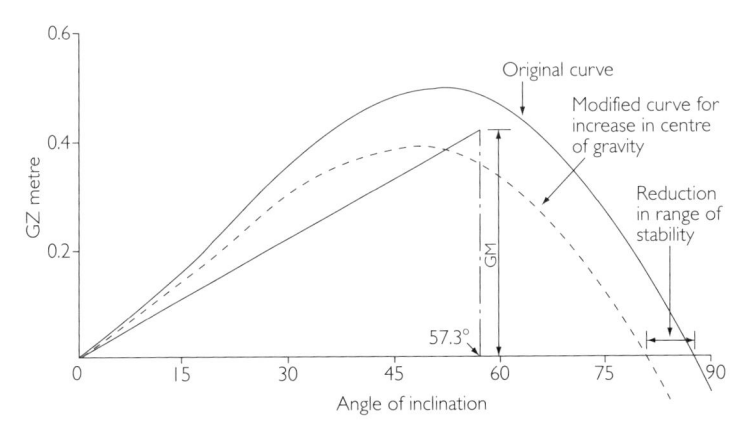

Fig 5.14: Curves of statical stability

Cross curves of stability indicate for a ship the range of stability, the maximum value of GZ and the angle at which it occurs, see Fig 5.14. The first two are related to freeboard, the greater its value the greater the range of stability and the maximum value of GZ. A modified curve is also shown, to indicate the loss in range of stability where the centre of gravity is raised. The slope of the curve at the origin is also important, in terms of the GZ value at small angles of inclination, and this is related to the metacentric height, GM. Since $GZ = GM \sin\theta$, then the slope can be found by differentiation. The tangent to the curve of statical stability at the origin can be determined by erecting an ordinate of value GM at an angle of 57.3deg (1 radian) and constructing a line to the origin.

Example

The angles of inclination and corresponding righting lever for a ship at an assumed KG of 6.5m are:

Inclination deg	0	15	30	45	60	75	90
Righting lever m	0	0.11	0.36	0.58	0.38	-0.05	-0.60

In a particular loaded condition, the displacement mass is made up as follows:

Item	Mass t	VCG m
Lightship	4200	6.0
Cargo	9100	7.0
Fuel	1500	1.1
Stores	200	7.5

Plot the curve of statical stability for this loaded condition and determine the range of stability.

The actual position of the centre of gravity, KG_1, is first found by taking moments about the keel:

$(4200 \times 6.0) + (9100 \times 7.0) + (1500 \times 1.1) + (200 \times 7.5)$
$\qquad = (4200 + 9100 + 1500 + 200)KG_1$

$$\therefore KG_1 = \frac{25\,200 + 63\,700 + 1650 + 1500}{15\,000} = 6.14\text{m}$$

The actual righting lever values are:
$\qquad G_1Z = GZ + GG_1\sin\theta \quad$ where $GG1 = 6.5 - 6.14 = 0.36$m

The new GZ values for the various angles of inclination can be determined in tabular form:

Inclination, θ deg	$\sin\theta$	$GG_1\sin\theta$	GZ m	G_1Z m
0	0	0	0	0
15	0.259	0.034	0.11	0.144
30	0.500	0.180	0.36	0.540
45	0.707	0.255	0.58	0.835
60	0.866	0.312	0.38	0.692
75	0.966	0.348	-0.05	0.298
90	1.000	0.360	-0.60	-0.240

A graph of the new G_1Z values plotted against inclination can be drawn, from which the range of stability was found to be 83deg.

5.17 Dynamical stability

Dynamical stability is the amount of work that must be done on a ship to heel it to some particular angle. It is a more representative measure of stability, since it considers a dynamic rather than, as previously, a static problem.

It can be found from the product of the ship's displacement and the area under the curve of statical stability to the angle considered. Alternatively the displacement can be multiplied by the difference between the centres of gravity and buoyancy in the upright and inclined positions.

Example
Using the tabulated values of G_1Z from the earlier example on cross curves of stability, determine the dynamical stability of the vessel when inclined to 60deg.

Dynamical stability = displacement x area under statical stability curve to 60deg.

The area can be found by using Simpson's rule and ordinate heights from the redrawn G_1Z_1 graph.

Inclination deg	G_1Z m	Simpson's multiplier	Area product
0	0	1	0
15	0.144	4	0.576
30	0.540	2	1.080
45	0.835	4	3.340
60	0.692	1	0.692
		Product total =	5.688

Area under curve to 60deg $= \dfrac{15}{57.3} \times \dfrac{1}{3} \times 5.688 = 0.496$

Dynamical stability = 15 000 x 9.81 x 0.496 = 72 986.4kNm.

5.18 Trim

Trim is the difference between the forward and aft draughts of a ship. Where the draught aft is greater than the draught forward, the ship is said to **trim by the stern**. Where the draught forward is greater than the draught aft, the ship is said to **trim by the head**. Where both values are the same, the ship is on an **even keel**. The axis about which a ship changes trim is known as the Longitudinal Centre of Flotation, F. Since trim changes about F, then the draught at this point will be constant for any given displacement.

Trim is a longitudinal inclination of the ship, and stability can also be considered in this plane, as it was for transverse stability. The same principles apply and, since the second moment of area of the waterplane about an axis through its centroid is many times greater than the second moment of area about the centreline, the longitudinal metacentric height is always a very large value. It is impossible for a ship to be unstable when inclined in the fore and aft direction and, thus, longitudinal stability will not be considered further.

The changes in draught and, therefore, trim, following the addition or moving of weights on a ship are important, particularly in shallow water. If a weight is added to a ship at the centre of flotation, F, then an increase in draught will occur, but there will be no change in trim. If a small weight is added to a ship at a point some distance from F, then the ship will change trim about F. If a large weight is added to a ship, at a point some distance from F, then the draught will change considerably and also the position of F and other characteristics of the ship. The addition of a small weight will be considered first.

5.19 Effect of adding a small weight

In order to simplify calculations, the weight is considered to be added first at the centre of flotation. The increase in draught is found by dividing the mass, ie, weight ÷ 9.81, by the **tonnes per centimetre immersion** (TPC). TPC is a value calculated for a ship by multiplying the waterplane area, A_W, and the density of seawater in tonnes per cubic metre and dividing by 100. In SI units, the Newtons per metre value would be $A_W \times 1025 \times 9.81 = 10\,055 A_W$. Since the value varies with draught, care must be taken in its use.

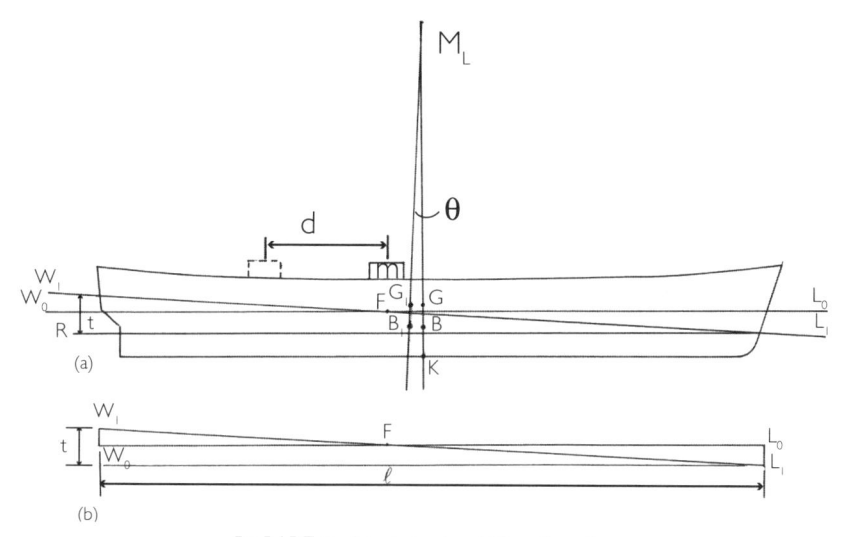

Fig 5.15: Trim change due to addition of small mass

The change in trim is calculated by considering the effect of moving the weight from the centre of flotation to its required position. Consider a ship of displacement, W, and length, L, at waterline W_0L_0 with a small weight, m, on the deck at the centre of flotation, F, see Fig 5.15(a). The ship's centre of gravity, G, is vertically in line with the centre of buoyancy, B. If the weight, m, is moved aft a distance, d, then the ship's centre of gravity will move aft from G to G_1, such that:

$$GG_1 = \frac{m \times d}{W}$$

The ship will then change trim about the centre of flotation, F, until it lies at some new waterline W_1L_1. The centre of buoyancy will then move from B to B_1, where B_1 and G_1 are vertically in line. The vertical through B_1 intersects the original vertical through B at M_L, the **longitudinal metacentre**. The distance, GM_L, is the **longitudinal metacentric height**.

$$GM_L = KB + BM_L - KG$$

$$BM_L = \frac{I_{yy}}{\nabla}$$

where I_{yy} is the second moment of area of the waterplane about a transverse axis through the centre of flotation, F, and ∇ is the volume of displacement.

If the vessel trims through an angle, θ, then
$$GG_1 = GM_L \tan\theta$$
and $\quad GM_L \tan\theta = \dfrac{m \times d}{W}$

thus $\quad \tan\theta = \dfrac{m \times d}{W \times GM_L}$

If a line RL_1 is drawn parallel to W_0L_0
 Change in trim = $W_1W_0 + L_0L_1 = W_1R = t/100$ m
where t is the change in trim in cm over length, L, m.

But $\quad \tan\theta = t/100L$

thus $\quad t = \dfrac{m \times d \times 100 \times L}{W \times GM_L}\quad$ cm

$m \times d$ is known as the trimming moment,

where trimming moment $= \dfrac{t \times W \times GM_L}{100L}\quad$ MNm

Now if t = 1cm, moment to change trim one centimetre (MCT 1cm) can be found since:

$$MCT\,1cm = \dfrac{W \times GM_L}{100L}\quad MNm$$

therefore:
change in trim, $\quad t = \dfrac{\text{trimming moment}}{MCT\,1cm}\quad$ cm

$\qquad\qquad\qquad\quad = \dfrac{m \times d}{MCT\,1cm}\quad$ cm

This change in trim will affect the end draughts. For the movement of weight considered, ie aft, the forward draught will be reduced and the after draught increased.

By similar triangles and reference to Fig 5.15(b), it can be seen that

$$\dfrac{t}{L} = \dfrac{L_0L_1}{FL_1} = \dfrac{W_1W_0}{W_0F}$$

The values t, L_0L_1 and W_1W_0 may be expressed in centimetres, while L, FL_0 and W_0F are expressed in metres.

Change in draught forward, $\quad L_0L_1 = \dfrac{-t}{L_0} \times FL_0\quad$ cm

Change in draught aft, $\quad W_1W_0 = \dfrac{+t}{L_0} \times W_0F\quad$ cm

Example
A ship of 10 000t displacement and 100m long, floats at draughts of 8.60m forward and 9.30m aft. The TPC is 12.1, GM_L is 110m and the centre of flotation, F, is 2.5m aft of midships.

Calculate: The MCT 1cm and the new end draughts when 92 tonnes are added 30m aft of midships.

$$\text{MCT 1cm} = \frac{W \times GM_L}{100L} = \frac{10\,000 \times 9.81 \times 110}{100 \times 100} = 1079.1 \text{MNm}$$

Bodily sinkage $= \dfrac{\text{mass}}{\text{TPC}} = \dfrac{92}{12.1} = 7.60\text{cm}$

Distance of mass from LCF = 30 - 2.5 = 27.5m
Trimming moment = 92 × 9.81 × 27.5 = 24 819.3 MNm

Change in trim $= \dfrac{92 \times 9.81 \times 27.5}{1079.1} = 23\text{cm}$ by the stern

Distance from F to fore end = (100/2) + 2.5 = 52.5m
Distance from F to aft end = (100/2) - 2.5 = 47.5m
Change in trim forward = (23/100) × 52.5 = 12.08cm
Change in trim aft = (23/100) × 47.5 = 10.92cm
New draught aft = 9.30 + 0.1092 = 9.41m
New draught forward = 8.60 - 0.1208 = 8.48m

5.20 Hydrostatic curves

Most of the quantities related to a ship will vary with draught, eg, displacement, position of the centre of buoyancy, centre of flotation and transverse metacentre. In order to simplify their determination, they are calculated for a range of waterplanes and plotted against a vertical axis of draught to become the **hydrostatic curves** for a ship, see Fig 5.16. Some of the values given are not in SI units and will need to be converted when used in calculations.

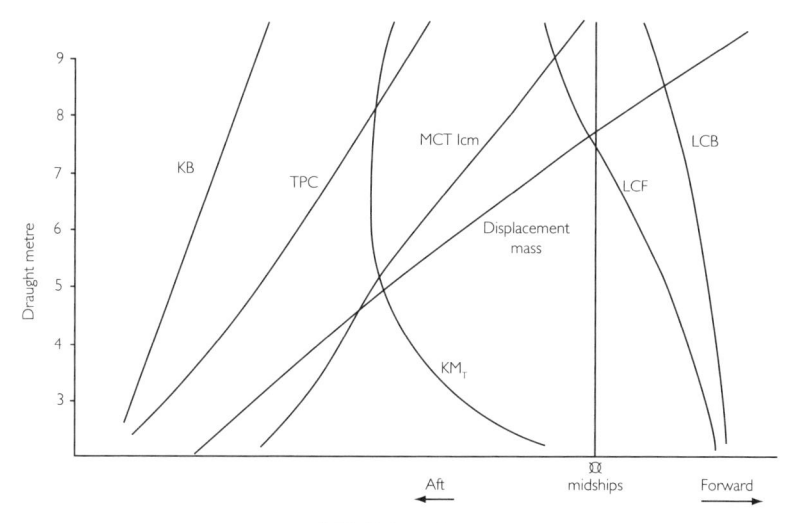

5.16: Hydrostatic curves

5.21 Effect of adding a large weight

When large weights are added to a ship the draught will increase considerably. Various other hydrostatic details will also change, making the calculation of the draughts more complicated. After loading (or discharging) any large weight, a ship will be in equilibrium. Thus the final centre of gravity and the final centre of buoyancy will be in the same vertical line.

If the end draughts are required, it will first be necessary to determine the position of the longitudinal centre of gravity and then the trimming moment acting. Hydrostatic data will be required for the vessel in order to determine the mean draught at the new displacement with the mass added. Values of MCT 1 cm and the positions of LCB and LCF from midships can then be found at the mean draught, assuming a level keel condition. It is unlikely that the LCG and LCB will be vertically in line and therefore a trimming moment will act. This will be the product of displacement and the horizontal distance between B and GM, and is known as the **trimming lever**.

Thus, total change in trim $= \dfrac{\text{trimming moment}}{\text{MCT1cm}}$

End draughts can be calculated by determining the change in trim forward and change in trim aft, as outlined earlier. When the vessel is floating at the new draughts, the centre of gravity G will be in line with the new centre of buoyancy, B_1.

Example
A ship 90m long has a light displacement mass of 1050t and LCG 4.60m aft of midships. The following items are loaded:

Cargo	2150t	LCG	4.70m forward of midships
Fuel	80t	LCG	32.60m aft of midships
Water	15t	LCG	32.90m aft of midships
Stores	5t	LCG	33.60m forward of midships

The following hydrostatic particulars are available:

Draught m	Displacement t	MCT 1 cm tonne-m	LCB from midships m	LCF from midships m
5.00	3533	43.10	1.00 forward	1.27 aft
4.50	3172	41.26	1.24 forward	0.84 aft

Calculate the final draughts of the loaded vessel.

Item	Weight MN	LCG	Moment forward MNm	Moment aft MNm
Lightweight	10 300.5	4.6 aft	-	47 382.3
Cargo	21 091.5	4.7 forward	99 130.05	
Fuel	784.8	32.6 aft		25 584.48
Water	147.15	32.9 aft		4841.24
Stores	49.05	33.6 forward	1648.08	
	32 373.0		100 778.13	77 808.02

Excess moment forward = 100 778.13 − 77 808.02 = 22 970.11MNm
LCG of loaded ship = 2341.5/3300 = 0.71m forward

The values of mean draught, MCT 1cm, LCB and LCF for the loaded ship can be found by interpolation from the given values.
Displacement mass difference 4.50 to 5.0m draught = 3533 − 3172 = 361t
Actual displacement difference for loaded ship = 3300 − 3172 = 128t
Actual draught difference = (128/361) × 0.50 = 0.1773m
Mean draught = 4.5 + 0.1773 = 4.6773m

MCT 1cm difference 4.50 to 5.0m draught = (43.10 − 41.26) × 9.81 = 18.05MNm
Actual MCT 1cm difference for loaded ship = (128/361) × 18.05 = 6.38MNm
MCT 1cm for loaded ship = (41.26 × 9.81) + 6.38 = 411.14MNm

LCB difference 4.50 to 5.0m draught = 1.00 − 1.24 = −0.24m
Actual LCB difference for loaded ship = (128/361) × -0.24 = −0.085m
LCB for loaded ship = 1.24 −0.085 = 1.16m forward of midships

LCF difference 4.50 to 5.0m draught = 1.27 − 0.84 = 0.43m
Actual LCF difference for loaded ship = (128/361) × 0.43 = 0.15m
LCF for loaded ship = 0.84 + 0.15 = 0.99m aft of midships

Trimming lever = 1.16 − 0.71 = 0.45m

Trimming moment = 3300 × 9.81 × 0.45 = 14567.85MNm by the stern

$$\text{Trim} = \frac{14\,567.85}{411.14} = 35.43\text{cm}$$

$$\text{Change in trim forward} = \frac{-35.43}{80}\left(\frac{80}{2}+0.99\right) = -18.15\text{cm}$$

$$\text{Change in trim aft} = \frac{+35.43}{80}\left(\frac{80}{2}-0.99\right) = +17.28\text{cm}$$

Draught forward = 4.6773 − 0.1815 = 4.4958m
Draught aft = 4.6773 + 0.1728 = 4.8501m

5.22 Displacement determination from measured draughts

If the displacement is required for a ship floating at draughts T_A and T_F, then a mean draught T_M can be found where:

$$T_M = \frac{T_A + T_F}{2}$$

An estimate of displacement can be made by reading from the hydrostatic curves at draught T_M. The hydrostatic curves are drawn for even keel waterlines and, thus, this value is inaccurate. The ship will float at waterline W_0L_0 for the draughts T_A and T_F and the value from the hydrostatic curves at draught T_M assumes a waterline W_1L_1 which intersects the original waterline W_0L_0 at midships. A ship trims about the centre of flotation, F, and, therefore, the level keel waterline for the same displacement as W_0L_0 is actually W_2L_2, see Fig 5.17.

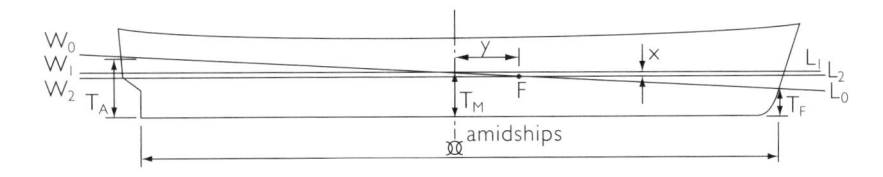

Fig 5.17: Displacement determination from measured draughts

If the thickness of the layer between the two waterlines is x and the distance from the centre of flotation, F, to midships is y, then,
thickness of layer, $x = (T/L) \times y$
where T is the trim, ie $T_A - T_F$

If the increase in displacement, per unit increase in draught, is λ, then,
displacement of layer $= (T/L) \times y \times \lambda$

The actual displacement at T_A and T_F is more accurately the value from the hydrostatic tables minus the displacement of the layer. Depending upon the position of the centre of flotation and the nature of the trim, ie by the head or the stern, the correction may be an addition or a subtraction. If trim and the centre of flotation are both forward or both aft, an addition is made, otherwise it is a deduction.

An additional correction may be needed if the ship is not floating in sea water of density $1025 kg/m^3$. No account has been taken of hogging or sagging in this estimation and, for a long vessel, a further correction may be necessary.

5.23 Effect of water density changes on draught

The displacement of a floating ship is equal to the mass of the volume of water it displaces.

Thus, displacement, W, = volume of displacement, $\nabla, \times \rho_{SW}$

The same ship when floating in fresh water would have the same displacement but would need to displace a greater volume of fresh water since the latter's density is less than that of sea water. The change in volume of displacement will be seen as an increase in draught, since the waterplane area A_W, is considered to remain constant.

Thus:
Change in volume of displacement $= (W/\rho_{FW} - W/\rho_{SW})$
Change in draught = change in volume of displacement/A_w
$\qquad = (W/\rho y_{FW} - W/\rho_{SW})/ A_W$

Now tonnes per cm immersion $= A_w \times density_{sw}/100$ for seawater,
thus $A_W = 100\ TPC_{SW}/density_{SW}$, and
change in draught $= W(\rho_{SW} - \rho_{FW})/(100\ TPC_{SW} \times \rho_{FW})$ m

A particular case of this effect occurs when a ship moves from sea water of density $1025 kg/m^3$ to fresh water of density $1000 kg/m^3$.

Change in draught $= W(1025 - 1000)/1000 \times TPC_{SW}$
$\qquad\qquad\qquad = 0.025 W/TPC_{SW}$ m, or $= W/40\ TPC_{SW}$ cm

This value is known as the **Fresh Water Allowance** and is used on the load line of a ship. It is the distance between the S line and the F line on the load line markings.

5.24 Docking and stability

When a ship is in a drydock from which the water is being pumped out, it will ultimately settle upon the blocks in the bottom of the dock. There will be a short time during which it is partly afloat and some of its weight is taken on the blocks. The conditions for stability at this time will be different to those for a floating vessel. There will also be a loading on the after end of the ship as it touches down.

A small trim by the stern is usual when entering a drydock, so the vessel will touch down on the blocks at the after end first. To determine this force on the aftermost block the ship can first be considered to be on an even keel and about to touch down on the blocks, see Fig 5.18.

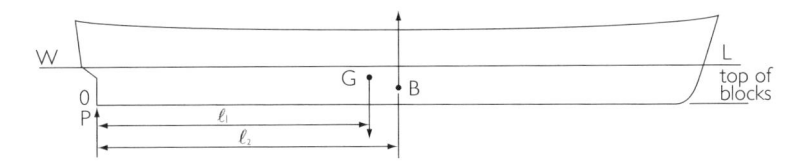

Fig 5.18: Ship touching down in drydock

If U is the buoyancy at the instant when the ship is about to make contact with the blocks all along the keel and W is the displacement of the ship when floating freely, then:

Force on aftermost block, $P = W - U$

If l_1 and l_2 are the distances of the centre of gravity, G, and the centre of buoyancy, B, forward of P, respectively, then:

$$Wl_1 = Ul_2 = (W - P)l_2$$

$$\therefore P = \frac{W(l_2 - l_1)}{l_2}$$

Now when trim by the stern is only small, the following approximation can be made:

$$Ul_2 = Wl_1 = \text{moment of displacement beneath waterline about O}$$
$$= Ul_2 + (W - U)a - Mt$$

where a is the distance of the centre of flotation of the even keel waterplane forward of the aftermost block, M is the moment to change trim 1cm and t is the trim in cm. Thus:

$$W - U = \frac{Mt}{a}$$

or, force on aftermost block, $P = \dfrac{Mt}{a}$

Stability is also a concern when a ship is lowered down onto blocks, since buoyancy is lost and a force acts on the aftermost block. The application of a force low down in the ship will tend to reduce stability. Consider the situation as shown in Fig 5.19, where a small angle of inclination, θ, is assumed. If moments are taken about the centre of gravity, G:

Righting moment $= (W - P)\,GM\sin\theta$,
Upsetting moment $= P \times KG\sin\theta$,
$\therefore$ net righting moment $= (W - P)GM\sin\theta - P.KG\sin\theta$
$= W\,GM\sin\theta - P.GM\sin\theta - P.KG\sin\theta$

$= \sin\theta(W.GM - P.GM - P.KG)$
$= \sin\theta[W.GM - P(GM + KG)]$

$$= W\sin\theta\left(GM - \frac{P}{W}KM\right)$$

The minus quantity within parentheses can be considered as a loss of metacentric height, due to the force P, at the keel. It must always be less than GM otherwise the ship will become unstable and could fall on its side.

Example
Just before touching down in a drydock, a ship of 5000t displacement mass floats at draughts of 2.7m forward and 4.2m aft. The length between perpendiculars is 150m and the water density 1025kg/m³. Using the given hydrostatic data, which may be considered constant over the variation in draughts considered, find:

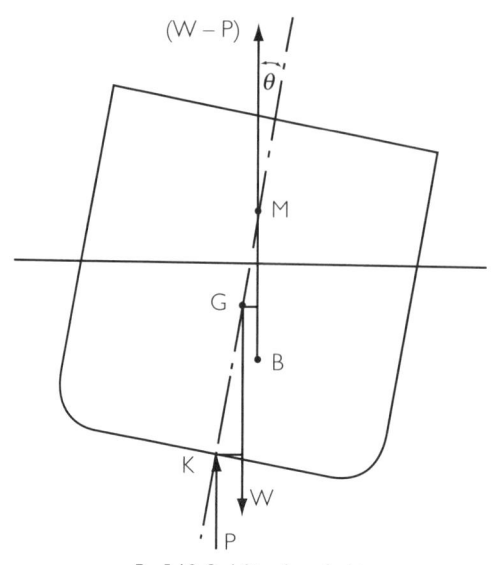

Fig 5.19: Stability when docking

(a) the thrust on the heel of the sternframe, which is at the after perpendicular, when the ship is just about to settle on the docking blocks, and,
(b) the metacentric height at the instant of settling on the blocks.

Hydrostatic data: KG = 8.5m, KM = 9.3m, MCT 1 cm = 1.05MN m, LCF = 2.7m aft of midships.

(a) Trim lost when touching down = 4.20 − 2.70 = 1.50m

Distance from heel of sternframe to LCF = (150/2) − 2.7 = 72.3m

Moment applied to ship when touching down = thrust on heel, P, × 72.3

Trimming moment lost by ship when touching down
= 1.50 × 100 × 1.05 = 157.5MNm

∴ thrust on heel, P = 157.5/72.3 = 2.18MN

(b) Loss of GM on touching down $= \dfrac{P}{W} \times KM$

$$= \dfrac{2.18 \times 1000 \times 1000 \times 9.3}{(5000 \times 1000 \times 9.81)N} = 0.41m$$

∴ metacentric height when touching down = 9.3 − 8.5 − 0.41 = 0.39 m

5.25 Squat
When a ship is underway it pushes its way through the water. The water which is pushed away will travel down the sides and under the keel of the ship. This moving water causes a dynamic effect, which brings about a drop in pressure beneath the hull, and the ship is drawn vertically down in the water. It will also trim until equilibrium is obtained. Squat is the mean increase in draught, ie sinkage, plus any con-

tribution due to trim, and is usually measured forward or aft, whichever is greater.

This phenomenon has become important in recent years due to the increasing size of tankers and bulk carriers, whose underkeel clearances are quite small in many ports. Also the increase in speed of vessels, such as containerships, has resulted in significant amounts of squat. The amount of squat occurring will be considerably increased in shallow water and, also, where the width of water available is limited. The speed of the ship and a number of hull features, such as draught, static trim and block coefficient, all contribute to the amount of squat. The ability to determine the amount of squat and an understanding of the influencing factors is, therefore, essential.

The main factor is the ship's speed, V_K. Squat will vary approximately with the square of the speed. The block coefficient, C_B, also has a direct relationship with squat. In addition, the value of C_B is used to assess whether the ship will trim by the head or the stern when squatting. If C_B is about 0.7 or greater, trim by the head or bow is most likely. Another important factor is the immersed cross-section of the ship's midship section, A_S, divided by the cross-section of the waterway, A_W, where this is a narrow channel. Using a procedure developed by Barrass[1], a 'blockage factor', S, is used to derive a 'velocity-return factor', S_2, where the two are related as follows:

$$S = \text{blockage factor} = \frac{A_W}{A_C} = \frac{\text{ship breadth} \times \text{draught}}{\text{canal breadth} \times \text{depth}}$$

$$S_2 = \text{velocity-return factor} = \frac{A_S}{A_W} = \frac{A_S}{A_C - A_S} = \frac{S}{1-S}$$

A formula for the estimation of squat in confined channels and in open water conditions is:

maximum squat, $\delta = \dfrac{C_B \times S_2^{2/3} \times V_K^{2.08}}{30}$ metres

where C_B is the block coefficient, S_2 is the velocity-return factor and V_K is the ship speed in knots. For maximum squat in open water, Barrass gives a simplified formula which provides satisfactory accuracy:

maximum squat, $\delta = \dfrac{C_B \times V_K^2}{100}$ metres

Other simplified formulae have been suggested by Dand[2]. These are:

1. 0.30m(1ft) for every 5kt of forward speed.
2. Squat equals 10% of the draught.
3. Use of the NMI squat chart[3].

Dand and Ferguson[4] have investigated squat by experiments on full-form ship models and also full-scale measurements on a number of ships. A semi-empirical method was developed to predict squat using one-dimensional theory. This was then modified on the basis of model tests and a computer program was developed to perform the calculations.

Full-scale measurements on a number of ships were then made which compared quite well with the calculated and model test results. In their conclusions they stated that squat would be noticeably affected by the initial trim of the ship, its speed and the water depth/static draught ratio. Also moderate, or negligible, effects might occur due to propeller loading and a block coefficient, C_B, in the range 0.82-0.90.

5.26 Flooding and subdivision

When the hull of a ship is pierced below the waterline, possibly by a collision or grounding, water will flood into the compartment created by the **subdivision bulkheads**. Buoyancy will be lost and draught will increase, as the ship sinks lower in the water. If the **reserve buoyancy** is greater than that lost, the vessel will not sink. Reserve buoyancy is the internal watertight volume of the compartments above the waterline and up to the bulkhead deck. The **bulkhead deck** is that uppermost deck to which the subdivision bulkheads extend. The subdivision bulkheads divide the ship into a number of watertight compartments along its length. Requirements for subdivision have been set by the International Convention on Safety of Life at Sea (SOLAS). Passenger ships must comply with certain standards of subdivision which are based upon a margin line, below which the vessel must not sink when flooded. The **margin line** is a line drawn parallel to, and 76mm below, the upper surface of the bulkhead deck at the ship's side. The loss of buoyancy and effects on stability as a result of flooding will be considered, before examining subdivision.

5.27 Direct flooding calculation

An intact ship initially floats at waterline, WL, with all compartments intact. When a forward compartment is flooded, the ship will now sink down to regain lost buoyancy, and the centre of buoyancy, B, will move aft, causing the ship to trim and take up some new waterline W_1L_1, with centre of buoyancy at B_1, see Fig 5.20. The **lost buoyancy** and the **added weight** methods are two approaches to the calculation of the new waterline, W_1L_1, and both involve an iterative approach.

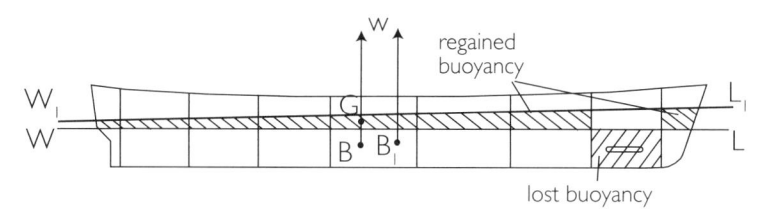

Fig 5.20: Flooding and buoyancy change

In any flooded or bilged space the volume of water able to enter will be less than its volume, owing to, at the very least, beams and frames or, much larger items of machinery and equipment. The percentage of a space which can be flooded is known as its **permeability**, μ, which can vary from about 95% for an empty compartment, to 60% or less for cargo holds.

When using the **lost buoyancy** method, the volume of the flooded compartment, to its original waterplane and the area of the waterplane which has been lost, are calculated, taking into account the permeability. If the ship is initially considered to sink parallel to its original even keel waterplane, then:

$$\text{parallel sinkage} = \frac{\mu v}{A - \mu a}$$

where μ is permeability, μv is the lost volume of buoyancy, A is the intact waterplane area and μa is the area lost after flooding.

A second approximation will be required for the sinkage, since the increase in draught will bring about a change in the area of the waterplane. This can be made for a waterplane at half the estimated sinkage, ie s/2. The area of the mean waterplane, A_M, and its second moment of area, I_M, about a transverse axis through its centroid must be calculated. The centroid of the waterplane is considered as the centre of buoyancy of the added layer. Thus:

$$\text{sinkage} = \frac{\mu v}{A_M}$$

The moment to change trim 1m for the mean waterplane, MCT_M must be calculated, where:

$$MCT_M = \frac{wGM_L}{L}, \text{ where } GM_L = KB + BM_L - KG, \text{ and } BM_L = \frac{I_M}{V}$$

where w is displacement, GM_L is longitudinal metacentric height, L is length and ∇ is volume of displacement. Thus:

$$\text{Trim} = \frac{\mu\rho v \bar{x}}{MCT_M}$$

where $\bar{x}$ is the distance between the centroid of lost volume and the centre of buoyancy of the added layer.

Now, if the distance of the longitudinal centre of flotation from amidships is γ, then

$$\text{Draught aft} = T + \frac{\mu v}{A_M} - \frac{(L/2 - \gamma)}{L} \times \frac{\mu\rho g \bar{x}}{MCT_M}$$

$$\text{Draught forward} = T + \frac{\mu v}{A_M} + \frac{(L/2 - \gamma)}{L} \times \frac{\mu\rho g \bar{x}}{MCT_M}$$

The sign convention adopted assumes the longitudinal centre of flotation is aft of midships and flooding takes place forward. Further approximations can be made if required. The position of the centre of gravity is considered unchanged in the lost buoyancy method.

In the **added weight method**, the water entering the flooded space is considered as an added weight and the centre of gravity of the ship must be recalculated. Due allowance must again be made for permeability and the free surface effect of the entered water, but all hydrostatic data used is that for the intact ship. An added weight-type calculation can be done and the new draughts calculated. The calculation must then be repeated to take into account the additional water that would enter at the new waterplane, until an acceptable level of accuracy has been achieved.

Example
A vessel of constant rectangular cross-section is 60m long and 10m wide. It floats at a level keel draught of 3m and had a centre of gravity 2.5m above the keel. Determine the fore and aft draughts if an empty, full-width, fore end compartment 8m long is flooded.

Lost buoyancy method

Area of intact waterplane, $A = (60 - 8) \times 10 = 520 \text{ m}^2$
Volume of lost buoyancy, $v = 8 \times 10 \times 3 = 240 \text{ m}^3$

Parallel sinkage, $s = \dfrac{240}{520} = 0.46\text{m}$

The vessel will now trim about the new centre of flotation, F_1. The position of F_1 can be found by taking moments about midships:

$(60 \times 10 \times 0) - (8 \times 10 \times (30 - [8/2])) = ((60 \times 10) - (8 \times 10)) F_1$
ie, $(8 \times 10 \times 26) = ((60 \times 10) - (8 \times 10))F_1$

$2080 = 520 F1$,

$\therefore F1 = -4$m or 4m aft.

$\text{MCT 1 m} = \dfrac{W \times GM_L}{L}$.

$KB_1 = \dfrac{T_1}{2} = \dfrac{3 + 0.46}{2} = 1.73$m.

$BM_L = \dfrac{I_L}{\nabla} = \dfrac{1/12(52)^3 \times 10}{60 \times 10 \times 3} = 65.1$m

$KG = 2.5$m (constant)

$\therefore GM_L = 1.73 + 65.1 - 2.5 = 64.33$m

$\therefore \text{MCT 1m} = \dfrac{60 \times 10 \times 3 \times 1.025 \times 9.81 \times 64.33}{60}$

$= 19\,405.6$ kNm

$\text{Trim} = \dfrac{\rho g v \bar{x}}{\text{MCT1m}} = \dfrac{1.025 \times 9.81 \times 240 \times 30}{19\,405.6} = 3.73$m

Draught aft $= 3 + 0.46 - \left(\dfrac{26}{60} \times 3.73\right) = 1.84$m

Draught forward $= 3 + 0.46 + \left(\dfrac{34}{60} \times 3.73\right) = 5.57$m

Added mass method

Mass added at 3m draught $= 3 \times 8 \times 10 \times 1.025 = 246$t

Parallel sinkage, $s = \dfrac{246}{1.025 \times 60 \times 10} = 0.4$m

New displacement mass $= 60 \times 10 \times 3.4 \times 1.025$
$= 2091$t.

$\text{MCT 1m} = \dfrac{wGM_L}{L}$

$KB_1 = \dfrac{3 + 0.4}{2} = 1.7$m

$BM_1 = \dfrac{I_L}{\nabla} = \dfrac{1/12(60)^3 \times 10}{60 \times 10 \times 3.4} = 88.2$m

The new centre of gravity, KG_1, can be found by taking moments about the keel:

$(60 \times 10 \times 3 \times 1.025 \times 2.5) + (246 \times 1.5) = 4981.5 KG_1$

$4981.5 = 2091 KG_1$

$\therefore KG_1 = 2.38m$

$$\therefore MCT\ 1m = \frac{2091 \times 9.81 \times (1.7 + 88.2 - 2.38)}{60}$$
$$= 29\ 933.3\ kNm.$$

$\text{Trim} = \dfrac{246 \times 9.81 \times 26}{29\,933.3} = 2.1m$

$\text{Draught aft} = 3 + 0.4 - \dfrac{2.1}{2} = 2.35m$

$\text{Draught forward} = 3 + 0.4 + \dfrac{2.1}{2} = 4.45m$

A second calculation considering the mass of water entering at 4.45m draught will give draughts forward and aft of 5.14m and 2.05m respectively. Third and further calculations would come progressively closer to the lost volume draught values.

5.28 Stability after flooding

When a compartment is flooded, its waterplane area is considered lost and the transverse moment of inertia, I_{xx} of the original waterplane will be reduced. Now since $BM = I_{xx}/V$, then the value of BM in the flooded condition will reduce. The ship will sink to some new waterline until the volume of displacement, V, is achieved by the remaining intact region of the ship. The height of the centre of buoyancy, KB, will rise because of the increased draught, while the height of the centre of gravity, KG, will remain the same. This combination of events will generally result in a reduction in metacentric height, GM.

Flooding of an off-centre compartment, such as a wing tank on a tanker, will bring about listing, in addition to sinkage and trim. The flooded ship will list until the centre of buoyancy of the intact hull is vertically in line with the centre of gravity. If the freeboard is small, then equilibrium may not be achieved before the deck edge immerses and the ship will capsize.

5.29 Floodable length

The effects of flooding on compartments of known size have so far been considered. During the initial stages of ship design the length of the vessel must be subdivided into compartments based upon **floodable length**. This is the length of a compartment in the region considered that can be symmetrically flooded without the ship sinking below the margin line. Symmetrical flooding assumes no listing of the ship.

Bulkheads subdivide the length of a ship and determine the size of compartments. Classification societies require cargo ships to have a minimum number of bulkheads according to their length. These requirements are based on both transverse structural strength and flooding considerations. All ships must have a collision- or fore peak bulkhead, an after peak bulkhead and a bulkhead at each end of the machinery space. The after bulkhead for an aft-located machinery space can be the after peak bulkhead. Additional bulkheads are required according to the ship's length. Regulatory requirements for subdivision for passenger ships are set by the International Convention on Safety of Life at Sea (SOLAS). Some countries, eg United States of America, have national standards of subdivision which apply to cargo ships.

These standards require the naval architect to assume some extent of damage, eg, size of opening, location, such as between or on a bulkhead, and the acceptable conditions of flotation and stability after flooding, margin line not submerged. The determination of floodable length will first be outlined, and followed by the different standards which can be applied to satisfy the regulations.

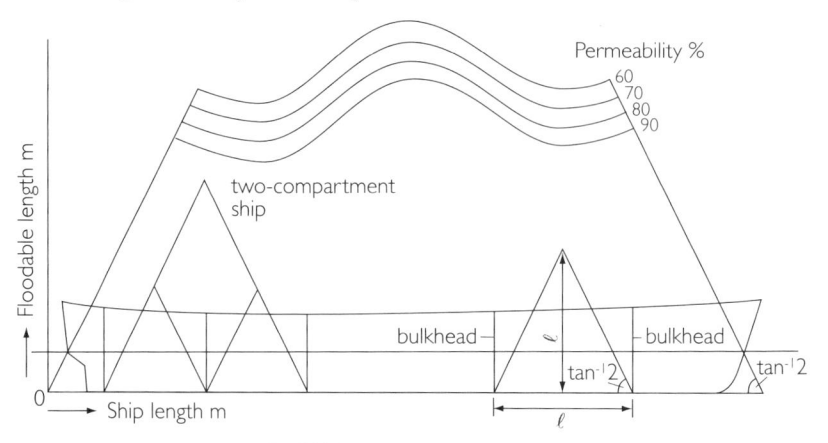

A floodable length curve can be created for a ship, based upon a waterline tangential to the margin line, Bonjean curves, and sectional area curves to the trimmed waterlines. A trial floodable length and centroid of the water volume entering are used in a numerical integration of the trim line sectional area curve, to determine the volume of flooding water and the centroid and mid-position of the flooded compartment which results. Further iterations will define the compartment more precisely. Floodable length will vary according to position along the ship's length and also the permeability of the space considered, see Fig 5.21.

The curves are used to determine suitable locations for subdivision bulkheads by drawing isosceles triangles, as shown in Fig 5.21. The fore and aft peak bulkheads will be located according to classification society rules, and machinery space and cargo hold bulkheads will be provisionally located. At each of the locations of the bulkheads an isosceles triangle is drawn, where the height is the distance between the bulkheads, l. If the vertex of the triangle is below the floodable length curve at the appropriate permeability, then the ship will not sink below the margin line due to flooding. The height of the triangle is the actual flooded length of compartment. Where it is some distance from the floodable length curve, the final trim line will be well below the margin line. This assessment ensures the vessel is a **one-compartment ship** for sinkage and trim. If the sides of the triangles are extended to form triangles enclosing a bulkhead and do not reach the floodable length curve, the vessel is said to be a **two-compartment ship**, see Fig 5.21. A two-compartment ship can survive flooding of two adjacent compartments.

Fig 5.21: Floodable length curves

5.30 Subdivision criteria

The International Maritime Organization's SOLAS convention and various national standards set out various requirements in relation to acceptable conditions of flotation and stability after flooding. A number of different approaches are used.

5.30.1 Factor of subdivision

One of the oldest approaches to subdivision is embodied in the UK Merchant Shipping Regulations which defines a **factor of subdivision**. This is a numerical value of unity or less, used to convert the floodable length into a **permissible length** of compartment. This approach aims to reduce the distance between subdivision bulkheads and thus improve the likelihood of flooding without the vessel sinking. A **criterion of service** numeral is used to describe the extent to which the ship is a passenger vessel. A high number indicates a large number of passengers are carried. The factor of subdivision gets smaller as the criterion of service numeral increases. A one-compartment ship will have a factor of subdivision between 1 and $1/2$, while a two-compartment ship will be between $1/2$ and $1/3$. This approach brings the bulkheads closer together, which should increase the possibility of the vessel remaining afloat if flooded, but it also increases the likelihood of a bulkhead being damaged and two compartments being flooded.

5.30.2 Compartmentation

Regulations using this system require the ship to survive flooding if an integer number of compartments are flooded at the same time. Ships will be considered as one-compartment, two-compartment, etc, as previously outlined. Different parts of the ship may have different levels of compartmentation, based upon their lower statical likelihood of being flooded. This system is the inverse of the factor of subdivision system.

5.30.3 Probability of survival

The SOLAS 1974 Convention permits this approach to be used for passenger ship subdivision, as an alternative to the factor of subdivision system. Studies of casualty data form the basis of the probability of survival approach, together with model tests, which simulated the behaviour of flooded ships in various sea conditions. Three probabilities were considered to affect subdivision and damage stability. These are the probability that the ship may be damaged, that the ship will survive the flooding, and also that for a particular location and extent of damage, in the event that the ship is damaged. The regulations define an **attained subdivision index**, based upon probabilities relating to damage length, position and the effects of freeboard, stability and list in the flooded condition. This index for the ship must be equal to, or greater than, a **required subdivision index**, which is based on the length of the ship and the number of persons carried. The greater the number of passengers carried, the higher is the subdivision index and the amount of subdivision required.

5.31 References

1. Barrass CB. *Ship squat*, Polytech International (1978).
2. Dand IW. *The physical causes of interaction and its effects*, Nautical Institute Conference on Shiphandling (1977).
3. NPL Ship Division, *Estimating the bow and stern sinkage of a ship underway in shallow water*, The Naval Architect (Jan 1973).
4. Dand IW, and Ferguson AM. *The squat of full ships in shallow water*, Trans Royal Inst of Naval Architects (1973).

6 Ships and the sea

The operating environment for ships is the sea and some understanding of wave action, and the extreme conditions which can occur, is required in order to design vessels which can satisfactorily operate therein. The elements constituting this environment are first outlined and then waves are examined in relation to their nature and energy. Ship response to the sea, in particular pitching and heaving, is then considered as is ship motion in waves. Finally, seaworthiness and the seakeeping properties of a ship are discussed.

6.1 Environmental elements

The marine environment presents one of the most severe situations that any structure and items of equipment must endure. A ship is subject to the actions of the sea and the air, and the many changes in the state of each resulting in winds that can vary from a breeze to a storm, and sea states from flat calm to waves with speeds of more than 30m/s (60 knots).

The total salt content of the sea is called salinity and varies according to water density and temperature. The accepted values for water density used by naval architects are 1.000 tonnes/m^3 for fresh water and 1.025 tonnes/m^3 for seawater. Seawater temperature can range from $-1.5°C$ in the Antarctic to as high as $34°C$ in the Red Sea. Air temperatures can be as low as $-50°C$ to as high as $50°C$.

Wind can have a direct effect upon a ship, creating forces which increase resistance and make manoeuvring difficult. Winds also create waves which have major influence on a ship moving in the water. The severity of waves will depend upon the wind speed or strength, its duration and the distance over which it acts (fetch). Waves which are generated by local winds are referred to as a sea, while those which have travelled out of their area of generation are called swell. Wind strength is classified according to the Beaufort Scale, which is based upon the appearance of the sea surface, see Table 6.1. Wind speed is measured at a standard height of 6m above sea level, since there is an interaction at the wind and sea surface.

Beaufort Scale number	Descriptive terms	Limits of wind speed	
		kts	m/s
0	Calm	less than 1	0.3
1	Light Air	1-3	0.3-1.5
2	Light Breeze	4-6	1.6-3.3
3	Gentle Breeze	7-10	3.4-5.4
4	Moderate Breeze	11-16	5.5-7.9
5	Fresh Breeze	17-21	8.0-10.7
6	Strong Breeze	22-27	10.8-13.8
7	Near Gale	28-33	13.9-17.1
8	Gale	34-40	17.2-20.7
9	Strong Gale	41-47	20.8-24.4
10	Storm	48-55	24.5-28.4
11	Violent Storm	56-63	28.5-32.6
12	Hurricane	64 and over	32.7 and over

Table 6.1: Beaufort wind scale

6.2 Waves

Investigations into the motion of waves have resulted in their classification into four types:
1. Capillary waves or ripples.
2. Translation waves.
3. Oscillating or deep-sea waves.
4. Shallow water waves.

Capillary waves are of no concern to naval architects since they are of very short length and small height. A tidal wave is an example of a translation wave, where there is bodily movement of the fluid. Deep-sea waves have the greatest influence on a ship's motions and do not result in a bodily movement of the fluid, only an oscillation of water particles in a closed path. They are generally generated by the wind blowing across the sea surface. Gravity waves is another term used, since gravity will act against them to ultimately create smooth water. The irregular nature of the sea surface is usually analysed by considering it as made up of a number of regular systems, each of relatively small amplitude with a range of periods.

6.2.1 Regular waves

For a simplified approach, wave contours can be taken as sinusoidal. This enables straightforward mathematical analysis and is fairly realistic. Measurements of actual waves indicate they have sharper crests and broader troughs than sine waves. Some naval architectural work has used trochoidal wave shape, but this approach is mathematically more cumbersome and does not correctly represent some of the dynamic properties of ocean waves.

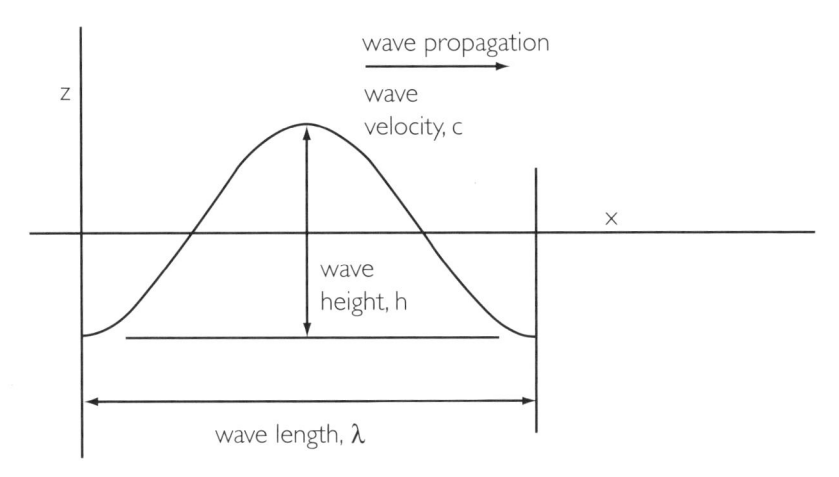

Fig 6.1: Sinusoidal wave form

Consider a wave travelling with velocity, c, along a horizontal plane, x, representing the still water level, see Fig 6.1. The surface wave height, z, at some point, x, and time, t, is:

$$z = \frac{h}{2}\sin(kx + \omega t)$$

where h is the wave height, k is the **wave number** ($2\pi/\lambda$), λ is the **wavelength**, ω is the **wave frequency** ($2\pi/T$), and T is the **wave period**.

The **wave velocity**, $c = \dfrac{\lambda}{T} = \dfrac{\omega}{k}$

Using either the sine or cosine function would produce correct forms of the above equations, it would merely indicate a different choice of origin. With the origin below a wave crest, the cosine function would be correct; with the origin at the nodal point before a crest, the sine function would be correct.

Thus:
$$z = \frac{h}{2}\sin(kx + \omega t)$$
$$z = \frac{h}{2}\cos(kx - \omega t)$$

The two parts of the cosine function can be used to define ocean waves. Considering the first part, $\cos kx = \cos 2\pi x/\lambda$, describes the spatial form of the wave at a specific time. When time, $t, = 0$ then wave height, z, would be:

$$z = \frac{h}{2}\cos\frac{2\pi x}{\lambda}$$

The second part of the function, $\cos\omega t$, describes the form of the wave at a specific point, as it varies with time. When $x = 0$, the wave height varies with time according to the expression:

$$z = \frac{h}{2}\cos\frac{2\pi t}{T}$$

This term describes the progressive nature of ocean waves, in that they travel along the water surface at a velocity, c, in a direction at right angles to the line of their crests. Note, it is the wave that travels and not the water mass. Water particle motion will occur involving vertical motion, since the wave crests are higher than the troughs. Water particle motion is actually orbital, with each particle on the surface following a circle of diameter equal to the wave height. As a wave crest passes a point, the surface particle will briefly move in the direction of wave propagation and as a trough passes, it will briefly move in the opposite direction. Between troughs and crests the particle will have both vertical and horizontal velocity components. Beneath the surface, particles will follow similar, but smaller, diameter circular orbits. These diameters will reduce exponentially with depth and at a water depth of more than half the wave length, the orbital motion is almost negligible

While water pressure is proportional to the distance below the surface, this is not the case in a wave. The pressure varies along the length of the wave, because the wave particles move in circular orbits. The resulting effect is that pressure is reduced below the static value at a wave crest and is increased at the trough. This will affect the buoyancy of a ship passing through waves and is known as the **Smith effect**.

The wave period, T, is the time interval between two crests passing a particular point and it can be found from the wave velocity, c, thus:

$$T = \frac{\lambda}{c} = \frac{\lambda}{\sqrt{(g\lambda)/2\pi}} = \sqrt{\frac{2\pi\lambda}{g}}$$

While this is the period relative to some fixed position, a ship will be moving through the sea and the **period of encounter**, T_e, relative to the ship will be required. If a ship is considered moving through a series of long-crested waves, of wave speed c_w, on a course inclined at some angle α, then the period of encounter will become $T_e(c_w - V_S\cos\alpha)$ where the ship speed is V_S, see Fig 6.2. This distance is equal to one wave length, thus:

$$\lambda = T_e(c_w - V_S\cos\alpha) \text{ or } T_e = \frac{\lambda}{c_w - V_S\cos\alpha}$$

Now, if the actual wave period, T_w, is considered then

$$T_w = \frac{\lambda}{c_w} \text{ and, thus}$$

$$T_e = \frac{\lambda}{c_w - V_s \cos\alpha} = \frac{\lambda/c_w}{1 - (V_s/c_w)\cos\alpha} = \frac{T_w}{1 - (V_s/c_w)\cos\alpha}$$

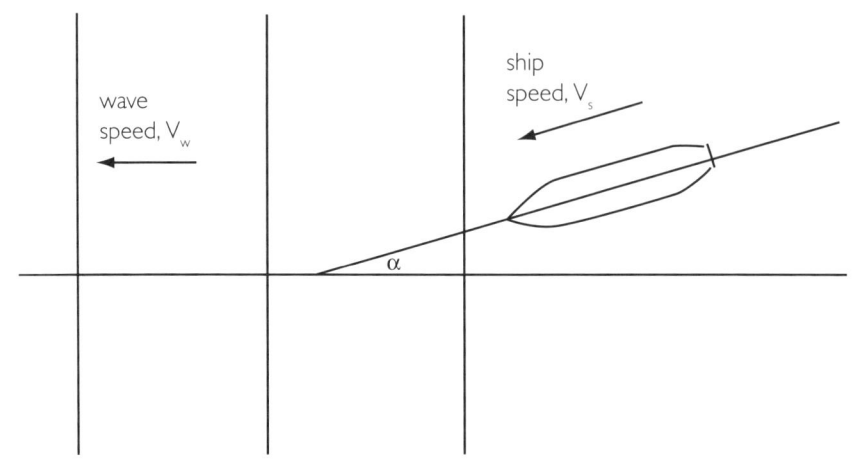

Fig 6.2: Period of encounter

If the ship is in a following sea, ie with the wave velocity in the same general direction as the ship, α is between zero and 90deg, then the period of encounter is greater than the wave period. When α is greater than 90deg, the period of encounter is less than the wave period.

6.2.2 Irregular waves

Any observation of actual sea conditions will indicate an irregular pattern of waves made up of components each with a particular length and height. At some distance from the point of their generation, waves may begin to show some indications of regularity. If, for the purposes of analysis, a series of waves was considered in which they were all travelling in the same direction and were of infinite width, then measurements could be made of wave heights and the distances between wave crests.

The apparent wave heights would be the vertical distance between adjacent crests and troughs and would vary along the waves. The time interval between two crests or a cycle of the wave movement would give the apparent wave period. In actual analysis of wave records the **significant wave height** is used which is the mean of the highest third of the waves. A sea state described as 'slight' would have significant wave height of 0.5 to 1.25m and 'very rough' would be 4 to 6m, according to one accepted code.

If actual wave height data was plotted in a histogram form using frequency of occurrence against wave height then a **Rayleigh distribution** occurs of the form

$$f(h) = \frac{2h}{E}\exp\left(\frac{-h^2}{E}\right)$$

where f(h) is the frequency of occurrence and h is wave height. The parameter $E = (1/N)\sum_1^N h^2$, and N is the total number of observations. The Rayleigh distribution is only relevant for short-term records of about 30min duration. Longer duration records will follow a **Normal** or **Gaussian** distribution.

6.2.3 Wave energy

The energy in a deep-sea wave can be shown to be half potential and half kinetic. For a sinusoidal wave, the total energy of a single wave per unit breadth is:

$$\text{wave energy} = \frac{1}{8}\rho g \lambda h^2$$

where ρ is the density of seawater, λ is the wave length and h is the wave height. Wave energy can be seen to be proportional to the wave length and the square of the wave height. An important conclusion from this expression is that the energy per unit area of sea surface is proportional to the wave height squared.

The total energy in an irregular sea can be represented by an energy spectrum. The wave patterns in the irregular sea can be represented by a series of regular components found by Fourier analysis. The elevation of the sea surface at any point in time is then:

$h = \Sigma h_n \cos(\omega_n t + \varepsilon(\omega_n))$

where h_n is the height and ω_n is the circular frequency of the nth component and $\varepsilon(\omega_n)$ is an arbitrary phase angle.

Now, since the energy per unit area of sea surface is proportional to the height squared, so the energy of a particular component will be proportional to h_n^2. The total energy of the sea will be the summation of the energies of all its components thus:

total energy $\propto \Sigma h_n^2$.

If the mean wave height over a small frequency range, $\delta\omega$, is found and then plotted at various values of ω, then a histogram of wave energy to a base of frequency will be obtained, see Fig 6.3. The ordinate is the **spectral density**, $S(\omega)$ which has the units of m^2s, since it represents the energy in a small interval of frequency whose unit is 1/s. The total energy of the waves will be proportional to $\int S(\omega)\delta\omega$.

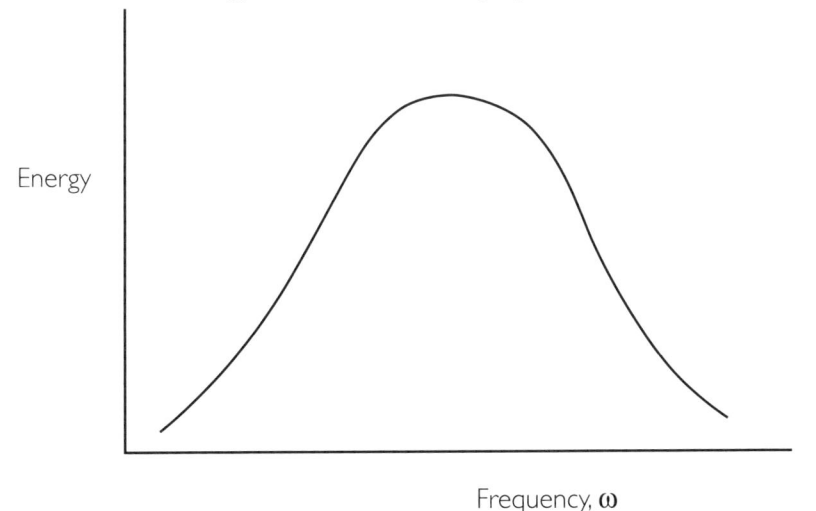

Fig 6.3: Energy spectrum

Wave spectra take on various shapes as a wave system develops and are dependent upon the wind speed, duration and fetch. In the initial stages of a wind creating waves, only the short wave length, higher frequency components are created and over time the spectrum height will rise and longer wave length, lower frequency, waves will appear. Similar effects result from increasing wind speeds. The lower ordinate values represent partially developed seas, which lead ultimately to a fully developed sea with the maximum ordinate height.

In the spectra discussed a unidirectional sea has been assumed. To more accurately represent actual sea conditions a spreading function, $f(\theta)$, would be introduced, where θ is the angle of the direction of a wave component to the dominant direction of the waves. Wave energy would then be a function of θ and ω, ie $S(\omega) \times f(\theta)$, and the total energy would be found by integrating this product.

Wave statistics data is now collected by satellite and has been published in tables for sea areas which include all the major sea trading routes used by merchant ships. The probability of occurrence of various sea conditions for the different seasons are thus readily available to naval architects to assess the conditions that a new ship design is likely to meet in service. This topic will be considered further in Chapter 7 when considering the probabilistic approach to structural strength.

6.3 Ship response to the sea

A ship at sea or lying in still water will be subject to a wide variety of stresses and strains resulting from forces acting from outside and within the structure. Forces from within the ship include the structural weight, cargo, machinery weight and the effects of operating machinery. External forces include the hydrostatic pressure of the water on the hull and the action of the wind and waves. A ship is free to move with six degrees of freedom producing three linear movements – **heave**, **surge** and **sway** and three oscillatory – **roll, pitch** and **yaw**, see Fig 6.4. The oscillatory movements are most significant and will now be considered in detail. The flexible nature of a ship's structure results in other degrees of freedom, which will be discussed later when considering strength and vibration.

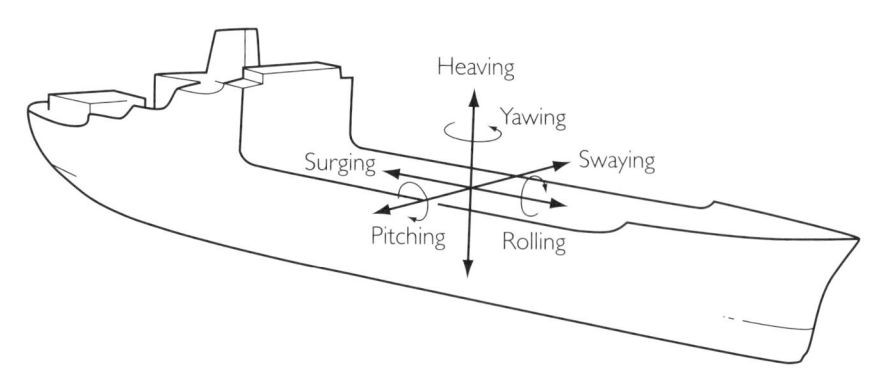

Figure 6.4: Ship movement – the six degrees of freedom

6.3.1 Oscillatory ship motions

Oscillatory ship motions can, if excessive, increase stresses in the structure and bring discomfort to both passengers and crew. Rolling and pitching tend to be the dominant oscillatory motions. It is important when considering these ship motions to determine the natural period of oscillation. The undamped motions of rolling, pitching and heaving will first be considered for a ship in still water, and then the effects of damping. Consideration will then be given to motion in waves.

Merchant Ship Naval Architecture

6.3.2 Rolling

If a ship at rest in still water is heeled to an angle, θ, by an external force, then a moment will act to return it to the vertical or upright position. The moment will be equal to MgGZ where GZ is the righting lever and M is the mass of the ship. If small angles of inclination are considered, then $GZ = GM\theta$. Using Newton's Law where force = mass × acceleration, the equation for the motion of the ship will be:

$$-MgGM\theta = Mk^2 \frac{d^2\theta}{dt^2}, \text{ since the moment of inertia of the ship, } I = Mk^2$$

thus $\frac{d^2\theta}{dt^2} + \frac{gGM\theta}{k^2} = 0$

This is the equation for simple harmonic motion, which has the solution:

$$\theta = A \sin\left(t\sqrt{\left(\frac{k^2}{gGM}\right)} + \delta\right)$$

the rolling period, T_r is given by:

$$T_r = 2\pi \frac{k_t}{\sqrt{gGM}}$$

k_t is the polar radius of gyration (approximately breadth/3)

Since the period of roll, T_r, is inversely proportional to the metacentric height, GM, then GM should be kept small for maximum human comfort, ie small accelerations and a long period. The addition or removal of masses, such as when handling cargo, will alter both k_t and GM. The period of roll is independent of the angle, θ, for angles up to about 10deg and such rolling is said to be **isochronous**. Where the period of roll is short a ship is said to be **stiff**, where it is long the ship is said to be **tender**.

6.3.3 Pitching

A similar expression can be obtained for the pitching of a ship, if the movement of water as the ship oscillates is neglected.

The pitching period, T_p, is given by:

$$T_p = \frac{2K_1}{\sqrt{GM_1}}$$

where K_1 is the longitudinal radius of gyration in metres about the transverse axis of pitching, which is generally through the ship's centre of gravity, G (K_1 = length/3 approximately).

The pitching period for a ship is about one-half to one-third of the rolling period.

6.3.4 Heaving

Heaving is the result of a downward vertical motion of a ship, z, and the resulting upward buoyancy force of $\rho g A_W \times z$. The equation of this motion gives a heaving period T_h which is given by:

$$T_h = 2\sqrt{\frac{\nabla}{A_w}} \qquad \text{where } \nabla \text{ is the volume of displacement.}$$

The periods of pitching and heaving are similar.

6.3.5 Cross coupling of motions

The oscillatory motions of a ship have been considered to act independently, but they are actually coupled. If heaving were to occur, even without an external pitching moment, pitching would still occur and vice versa.

Consider a ship moving down into a wave during heaving. Additional buoyancy will be created and can be considered added at the centre of flotation of the still waterline. Since the fore and aft form of the ship is not symmetrical, the centre of buoyancy will be at some distance from the centre of flotation. The centre of gravity will be unchanged and thus a moment will exist, tending to rotate the ship. The heaving motion will thus bring about a coupled pitching motion.

Example

A ship has a length of 97.5m; breadth 10.65m; draught 3.35m; transverse metacentric height = 0.76m; longitudinal metacentric height = 110m.
$K_t = 3.65m$; $K_l = 24.4$; $C_B = 0.54$; $C_W = 0.70$

Determine the natural periods of
1. Rolling.
2. Pitching.
3. Heaving.

1. $$T_r = \frac{2.0 K_t}{\sqrt{GM_t}} = \frac{2.0 \times 3.65}{\sqrt{0.76}} = 8.4s$$

2. $$T_p = \frac{2.0 K_l}{\sqrt{GM_l}} = \frac{2.0 \times 24.4}{\sqrt{110}} = 4.65s$$

3. $$\nabla = 97.5 \times 10.65 \times 3.35 \times 0.54 = 1878.42 m^3$$
$$A = 97.5 \times 10.65 \times 0.70 = 726.86 m^2$$

$$T_h = 2\sqrt{\frac{\nabla}{A}} = 2\sqrt{\frac{1878.42}{726.86}} = 3.2s$$

6.3.6 Added mass and damping

The oscillatory motions of the ship can be affected by the added mass of the water, which clings to the hull as it moves, and damping, caused by the viscous nature or the waves generated by the water. The water around a ship's hull is disturbed during its oscillatory motions and the result is an increase in the mass and inertia of the ship. The effect on rolling is to increase the radius of gyration by up to 5% while during heaving there may be an apparent doubling of the mass of the ship.

Since water is a viscous fluid, a moving ship experiences a resistance to motion. The energy used to overcome this resistance is called damping. If rolling is considered and the damping motion can be assumed to vary linearly with the angular velocity, then a differential equation can be produced for the damped motion. The period of damped motion can then be found. When damping is not proportional to the angular or linear velocity the differential equation is difficult to solve.

6.3.7 Motion in waves

In order to examine the motion of a ship in waves, both regular and irregular waves must be considered. Regular seas can be readily represented by combining a large number of sinusoidal waves. A ship's response to such a sea can, likewise, be considered the combination of responses to the individual sinusoidal waves.

A number of assumptions are made when studying a ship's response to regular waves. The pressure distribution in the wave is assumed to be unaffected by the ship's presence, which enables the motion to be considered as that for still water with the addition of a forcing function. The force is the summation of pressures acting on the immersed hull due to the wave.

In considering the different motions already outlined, heaving will take place due to the wave forces acting on the hull. The extent of the heaving motion will be greatest when the period of encounter with the waves is close to the ship's natural period of motion when heaving. If the ratio of the heaving period to the encounter period is used the maximum amplitude of motion will occur as the ratio approaches unity, ie resonance.

The only practical way to avoid this situation is for the ship to change course and thus change the period of encounter. Pitching and rolling can be considered in much the same way.

The amplitudes of the motions discussed are important when comparing different ship designs. The amplitude of heave will, for instance, vary with the height of the waves, since the heaving force can be assumed to be proportional to wave height. If heaving amplitude divided by wave height is used, this non-dimensional value is known as a **response amplitude operator**. A plot of the response amplitude operator against the ratio of the heaving period to the encounter period shows the maximum value occurring at unity, see Fig 6.5. The effect of damping limits the value of the amplitude, preventing it becoming infinite. Data for such information is usually obtained by running model tests in regular waves.

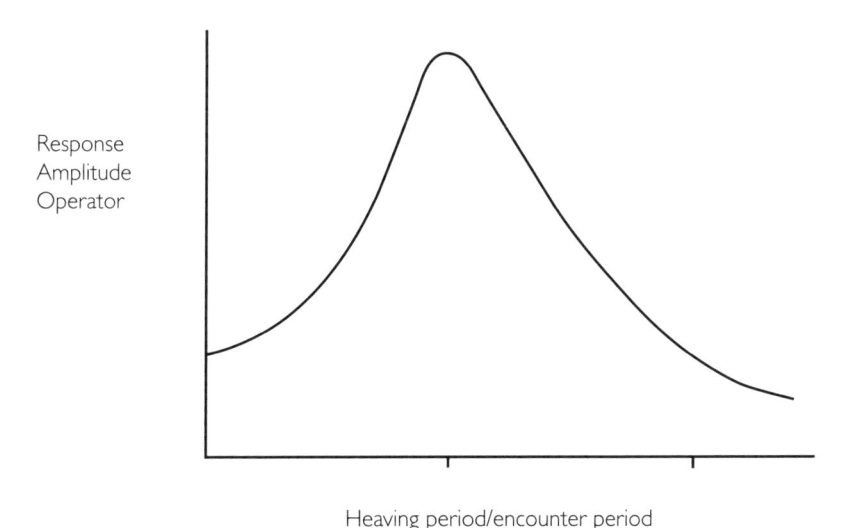

Fig 6.5: Response amplitude operator

Irregular seas can also be analysed by considering them as a series of regular components with varying frequencies and amplitudes. Each component could then be considered in relation to the response amplitude operator at the particular frequency and then multiplied by the wave height to obtain the actual amplitude of the motion. The frequency considered will be the frequency of encounter for the vessel with the waves. The energy spectrum for the ship would then be the

summation of all these responses.
Thus:

$$\text{motion energy spectrum, } S_m = \int_0^\infty S(\omega)(RAO)^2 \, d\omega$$

where $S(\omega)$ is the ordinate of the sea spectrum (ie, the wave height squared).

Once the motion energy spectrum has been found, then various values of motion amplitude, eg, average and significant, can be found. If more than one type of sea were to be considered, the data would have to be modified to take into account different weather intensities and the probable time a ship would be present in such weather intensities. This would ultimately lead to a probability of any particular motion occurring over the assumed lifetime for the ship design considered.

This method of analysis will be discussed in more detail in Chapter 7 when considering the probabilistic approach to ship structural strength.

6.4 Seaworthiness and seakeeping

A ship is designed to carry and deliver cargo in good condition and the naval architect must design the vessel to achieve this aim. Seaworthiness and seakeeping are terms used to describe aspects of this function. A vessel is said to be **seaworthy** if it is fit in all respects for the anticipated perils of the voyage and will carry the cargo and crew in a safe condition. It also refers to fitness to receive, carry, and preserve, the particular cargo. **Seakeeping** refers to the many aspects of a ship's design and construction which determine its ability to operate efficiently at sea, eg, stability, strength and speed. The effect of wave actions and the sea on the criteria affecting seaworthiness and seakeeping will now be considered.

6.4.1 Strength

The structure of a ship must have adequate strength to withstand the many forces and loads imposed upon it as a result of both static and dynamic effects. Ship strength will be discussed in detail in Chapter 7. Mention will be made at this point, of **slamming** which is the effect resulting from the rise and fall of the forward end of a ship when it is heaving and pitching. Large forces can act on the forward bottom plating of a ship. The term **pounding** is also used.

Additional structural strength must be provided from the forward perpendicular aft for 25-30% of the ship's length. The slamming action is short in duration but can set up vibrations in the structure which could lead to fatigue failure. A change in speed or direction can reduce or eliminate slamming.

6.4.2 Freeboard

A minimum height at the side of a ship is necessary to ensure satisfactory navigation and ship operation in addition to providing protection for the ship's cargo and equipment. This minimum height is part of the term freeboard. Freeboard is defined as the vertical distance from the summer load waterline to the top of the freeboard deck plating, measured at the ship's side amidships. The freeboard deck is normally the uppermost complete deck exposed to weather and sea, which has permanent means of closing all openings in the weather part and below which all openings in the side of the ship are fitted with permanent means of watertight closing. Freeboard was discussed in Chapter 3, with reference to the Load Line rules.

Adequate freeboard will reduce the amount of water reaching the deck of the ship as a result of heavy spray or large (green) seas. Deck erections, superstructures and sheer will also reduce wetness, as this problem is sometimes termed. Adequate arrangements to quickly drain or release this water from a ship's deck must be provided.

In addition to these design features, wetness can be lessened, by a reduction in speed or a change of course in relation to the waves.

6.4.3 Stability

Adequate stability is essential for seaworthiness, and appropriate loading and stowage of the cargo and any ballast should ensure this. The Load Line rules require that sufficient information is carried on board a ship to enable loading and ballasting to be achieved without creating any unacceptable stresses on the structure. No actual standard of stability is stated except for a ship in a damaged condition. Stability in this situation must be sufficient to enable the ship to withstand the final state of flooding of a compartment or compartments within the floodable length.

In broad terms the stability of a ship should be adequate when the metacentric height lies in the range 0.4 to 0.75m.

6.4.4 Ship operation

The speed of a ship is a significant factor in relation to many wave effects and thus a change in speed will often reduce such loadings on the ship. Wave effects can also simply slow down a ship when the effects upon the structure are not a problem and thus a greater engine output power is required to maintain the speed.

In certain wave conditions the propeller may partially emerge from the water resulting in a loss of ship speed and possible vibrations. A reduction in ship speed may be necessary to overcome this problem.

6.4.5 Seasickness

The motions of a ship can cause various forms of human discomfort referred to as seasickness. Rolling is usually the motion most likely to cause seasickness as a result of the acceleration experienced and its period. Frequencies of 0.15 to 0.2Hz appear to cause most problems. Where wave actions result in motions of this nature, action must be taken to avoid them by a change of speed or direction

6.4.6 Overall performance

In making an overall assessment of a ship's performance, or a comparative assessment between different designs, the designer must consider the various responses of a ship to different sea conditions experienced during its operating lifetime. Probabilities can be determined in relation to the meeting of various sea conditions, the loaded condition of the ship, the ship speed and direction, and any critical responses in relation to ship operations.

Wave data can be used and a theoretical approach to the calculation of probabilities, alternatively model testing or full-scale trials can be undertaken. Sea kindliness describes a ship which performs well in various sea conditions and weather. Factors which improve sea kindliness include increased length, increased size and higher freeboard. Bulbous bows can reduce some ship motions and a large flare at the bow may reduce wetness.

6.4.7 Motion control

There is a limit to the extent that amplitudes of motion of a ship can be reduced by changes in hull shape, but considerable reductions in roll amplitudes are possible by various forms of stabilisation. Rolling is the only ship motion that can be effectively dampened and reduced on ocean-going ships. Roll stabilisation systems can be either passive or active. In passive systems no external source of power nor special control system is required. Motion control is covered in Chapter 10.

7 Structural strength

A ship, whether at sea or lying in still water, is constantly being subjected to a wide variety of stresses and strains resulting from the action of forces from outside and within the ship. Internal forces result from the weight of the structure, the cargo and the machinery, together with the effects of operating machinery. External forces are created by the hydrostatic pressure of the water and the action of the wind and waves on the hull.

These forces are constantly varying in intensity and frequency but, for simplicity, will be considered individually, and by an initial classification into static and dynamic forces. Static forces are a result of the differences in weight and buoyancy occurring along the length of the ship. Dynamic forces result from the ship's motion in the sea and the action of wind and waves. A ship is free to move with six degrees of freedom, as mentioned in Chapter 6, resulting in linear and rotational movements. These static and dynamic forces create longitudinal, transverse and local stresses in the ship's structure. The greatest stresses occurring in the ship's structure are due to the distribution of loads along the ship, which cause longitudinal bending.

7.1 Longitudinal stresses

Static and dynamic loading of a structure will be considered separately in this examination of the longitudinal stresses.

7.1.1 Static loading

When a ship is floating in still water, two different forces will be acting along the length of the vessel. The weight of the ship, its cargo and machinery, will be acting vertically downwards. The buoyancy, or vertical component of hydrostatic pressure, will be acting upwards. In total the two forces exactly equal and balance one another, such that the ship floats at a particular draught. The centre of the buoyancy force and the centre of the weight will be vertically in line.

However, at various points along the ship's length there may be an excess of buoyancy or an excess of weight. Consider first the curve of buoyancy, which represents the upward force at various points along the length of the ship, see Fig 7.1 (a). The buoy-

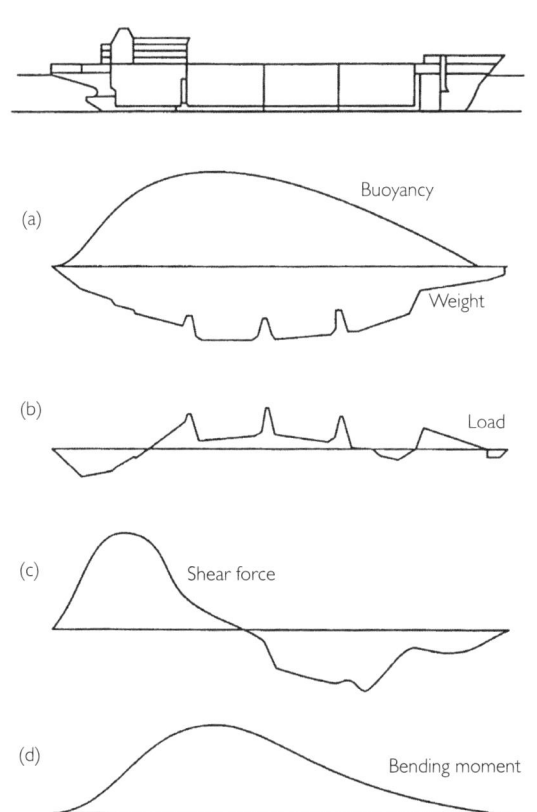

Fig 7.1: Static loading of a ship's structure

ancy forces increase from zero at the ends of the ship's waterline to a constant value over the parallel middle body section. The area within the curve represents the total upthrust or buoyancy exerted by the water.

Consider now the weight of the ship, which is made up of the steel structure, items of machinery, cargo, etc. The actual weight at various points along the length of the ship is unevenly distributed and is represented by a weight curve, as shown in Fig 7.1 (a). The weight curve actually finishes at the extremes of the ship's structure and can thus extend beyond the buoyancy curve.

At various points along the ship's length the weight may exceed the buoyancy, or vice versa. Where a difference occurs, this results in a load at that point. The load diagram, see Fig 7.1 (b), illustrates the actual loads at various points.

This loading of the ship's structure results in forces which act up or down and create shearing forces. The shear force at any point is the vertical force acting. It can also be considered as the total load acting on either side of the point or section considered. The actual shearing force at any section is, in effect, the area of the load diagram to the point considered. A shear force diagram can thus be drawn for the ship, see Fig 7.1 (c).

The loading of the ship's structure will also bring about bending. The bending moment at any point is the sum of the various moments to one side or the other. The bending moment at a section is also represented by the area of the shear force diagram to the point considered. A bending moment diagram is shown in Fig 7.1 (d). The maximum bending moment occurs when the shear force is zero.

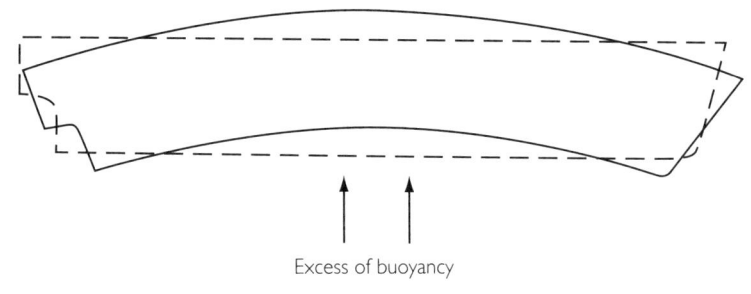

Fig 7.2: Hogging condition

Since a bending moment acts on the ship, then it will tend to bend along its length. This still water bending moment (SWBM) condition will cause the ship to take up one of two possible extreme conditions. If the buoyancy forces in the region of midships are greater than the weight, then the ship will curve upwards or 'hog', see Fig 7.2. If the weight amidships is greater than the buoyancy forces, then the ship will curve downwards or 'sag', see Fig 7.3.

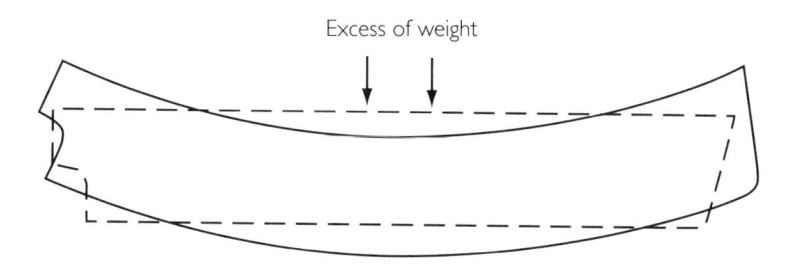

Fig 7.3: Sagging condition

7.1.2 Dynamic loading

If the ship is now considered to be moving through waves, the distribution of weight will remain the same. The distribution of buoyancy, however, will vary as a result of the waves. The movement of the ship will also introduce dynamic forces.

The traditional approach used to solve this problem is to convert the dynamic problem into an equivalent static one. To do this the ship is assumed to be balanced on a static wave the same length as the ship. The still water condition is shown in Fig 7.4(a).

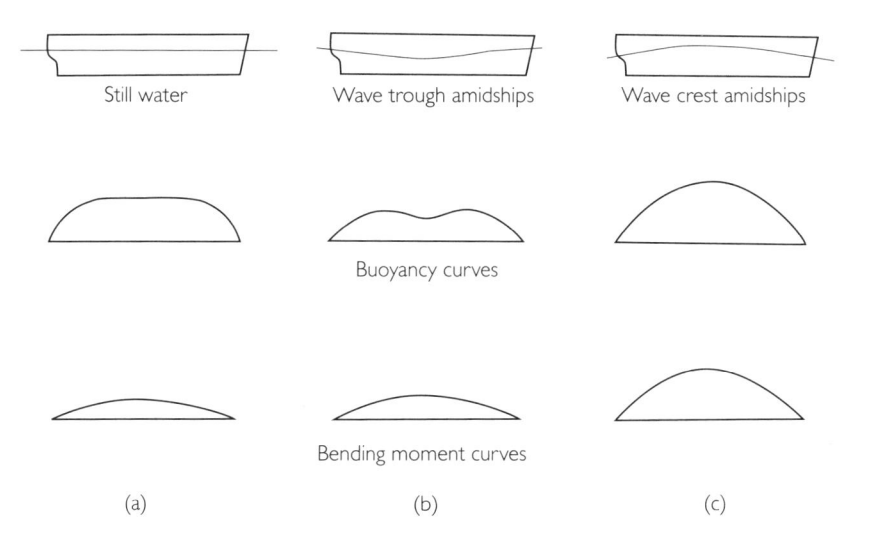

Fig 7.4: Dynamic loading of a ship's structure: (a) still water condition; (b) sagging condition; (c) hogging condition

If a wave with its trough at midships is now considered, then the buoyancy in this region will be reduced. With the wave crests positioned at the ends of the ship, the buoyancy here will be increased. This loading condition will result in a bending moment which will cause the ship to sag, see Fig 7.4(b). Since the ship in its still water condition is considered to hog, then this change to a sagging condition has required a bending moment to overcome the initial hogging bending moment in addition to creating sagging. The actual bending moment in this condition is therefore considerable and, it is an extreme condition.

If the wave crest is now considered at midships, then the buoyancy in this region will be increased. With the wave troughs positioned at the ends of the ship, the buoyancy here will be reduced. This loading condition will result in a significantly increased bending moment which will cause the ship to hog, see Fig 7.4 (c). This will, again, be an extreme condition giving the maximum bending moment that can occur in the ship's structure for this condition.

If actual loading conditions for the ship which will make the above conditions worse are considered, ie heavy loads amidships when the wave trough is amidships, then the maximum bending moments in normal operating service can be found. The ship's structure will be subjected to constantly fluctuating stresses resulting from these shear forces and bending moments as waves move along the ship's length.

7.2   Stressing of the ship's structure

Stresses are set up in the structure as a result of the bending of a ship. Sagging causes tensile stresses to be set up in the bottom shell plating and compressive stresses in the deck. When the ship hogs, tensile stresses occur in the deck and compressive stresses in the bottom shell. This stressing, whether compressive or tensile reduces in magnitude towards a position know as the neutral axis. The neutral axis in a ship is somewhere below half the depth and is, in effect, a horizontal line drawn through the centre of gravity of the ship's section. The fundamental bending equation for a beam is:

$$\frac{M}{I} = \frac{\sigma}{z}$$

where M is the bending moment, I is the second moment of area of the section about its neutral axis, σ is the stress at the outer fibres, and z is the distance from the neutral axis to the outer fibres.

This equation has been proved in full-scale tests to be applicable to the longitudinal bending of a ship. From the above equation:

$$\text{stress, } \sigma = \frac{M}{I/z}$$

This gives a value for the stress in the material at some distance, y, from the neutral axis. The values M, I and z can be determined from the ship, and the resulting stresses in either the deck or the bottom shell plating can be found. The ratio I/z is known as the section modulus, Z, when z is measured to the extreme edge of the section, ie the deck or the bottom shell plating. The values are determined for the midship section, since the greatest moment will occur at or near midships, see Fig 7.1.

The structural material included in the calculation for the second moment, I, will be all the longitudinal material which extends for a considerable proportion of the ship's length. This material will include side and bottom plating, inner bottom plating (where fitted), centre girders and decks. The material forms what is known as the hull girder, whose dimensions are very large compared with its thickness.

7.3   Calculating static longitudinal stresses

In performing what has become the standard calculation of static longitudinal stresses, the ship is assumed to be positioned upon a trochoidal wave of length equal to that of the ship. The value of this approach is that it determines the maximum bending moments the ship is likely to experience at sea. Wave height for the calculation is generally taken as $0.607(L)^{0.5}$, where L is in metres.

Two extreme conditions are considered, a wave crest amidships and wave crests at the ends of the ship. The first will create hogging while the second causes sagging. The total bending moment obtained is often split up into the still water and wave bending moments. This is because the still water bending moment is influenced by both the mass and the buoyancy distributions, whereas the wave bending moment is only dependent upon the geometry of the ship and the wave.

To actually consider the ship balanced on a wave, the centre of gravity and the centre of buoyancy must be vertically in line, which may require a number of attempts at positioning the wave vertically on the ship's side. One method of achieving this is to draw the wave profile and superimpose this over a profile of the ship on which the Bonjean curve has been drawn in at each ordinate. A trial wave position and one ordinate are shown in Fig 7.5. The immersed areas at each ordinate can be read for the Bonjean curves where the wave profile cuts the ordinate. These areas are then put through Simpson's Rule to give the displacement and the position of the LCB. The position of the wave that ensures the displacement equals the weight, and that the centre of gravity and centre of buoyancy are vertically in line, can be found by trial and error.

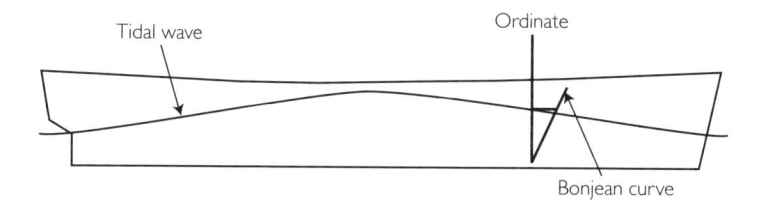

Fig 7.5: Balancing a ship on a wave

The buoyancy or upward force on a one-metre length of ship is the product of the density of the water (ρ) and the immersed volume (A) of the section considered. A curve of buoyancy per unit lengths can thus be drawn to a base of the length of the ship. The total buoyancy will be represented by the area under the curve, and the fore and aft centroid of the area represents the longitudinal centre of buoyancy.

The weight curve for a ship will require a knowledge of the various weights (m × g) along the length of the ship for the particular condition of loading considered. Once established and drawn, probably with the use of a computer program, the area under the curve will represent the displacement, and the fore and aft centroid represents the longitudinal centre of gravity.

The load curve is established from the difference between the weight per unit length and the buoyancy per unit length at the various points along the ship's length. The areas above and below the base line must be equal since the ship is in static balance.

The shearing force curve can be found by integrating the load curve:
shear force (SF) = $\int(\rho g A - (m \times g))dx$

The bending moment curve is found by integrating the shear force curve:
bending moment = $\int(SF)dx$

A tabular method of integration can be used for this type of calculation.

Example

A ship of length 130m, when poised on a wave, has the following mean values of weight and buoyancy as measured at the mid-point of each of 10 equally-spaced stations. Determine the shear force and bending moment at midships.

Station	0	½	1½	2½	3½	4½	5½	6½	7½	8½	9½	10
Weight Tonnes/metre	0	75.5	111.0	126.1	125.3	88.8	106.7	96.8	120.8	124.4	103.9	0
Buoyancy Tonnes/metre	0	53.3	93.3	115.5	125.3	131.5	133.3	133.3	123.3	105.7	64.9	0

The calculation can be done by setting up a table:

Station	Weight	Buoyancy	Load	SF	SF areas	BM areas
0	-	-	-	0	-	0
½	75.5	53.3	-22.2	-	-22.2	-
1	-	-	-	-22.2	-	22.2
1½	111.0	93.3	-17.7	-	-62.1	-
2	-	-	-	-39.9	-	84.3
2½	126.1	115.5	-10.6	-	-90.4	-
3	-	-	-	-50.5	-	174.7
3½	125.3	125.3	0	-	-101	-
4	-	-	-	-50.5	-	275.7
4½	88.8	131.5	42.7	-	-58.3	-
5	-	-	-	-7.8	-	334.0
5½	106.7	133.3	26.6	-	11.0	-
6	-	-	-	18.8	-	323
6½	96.8	133.3	36.5	-	74.1	-
7	-	-	-	55.3	-	249
7½	120.8	123.3	2.5	-	113.1	-
8	-	-	-	57.8	-	136
8½	124.4	105.7	-18.7	-	96.9	-
9	-	-	-	39	-	39
9½	103.9	64.9	-39	-	39	-
10	-	-	-	0	-	0

Bending moment at midships = 334.0 × 13 × 13 × ½
= 28 220tm = 276.9MNm

Shearing force at midships = 7.8 × 13 = 102t = 1.02MN

7.3.1 Characteristics of shear force and bending moment curves
A number of relationships have already been mentioned with respect to shear force and bending moment curves and are summarised as follows:

- The area under the weight curve and the buoyancy curves are equal.
- The centroids of the areas of weight and buoyancy are in the same athwartship section.

Merchant Ship Naval Architecture

- In the load curve the areas above and below the base line are equal.
- The maximum value of the shear force occurs where the load curve crosses the base line.
- The maximum bending moment occurs where the shear force curve crosses the base line.
- The shearing force and bending moment curves must have zero values at their ends.

For any ship of fixed displacement at particular still water draughts, the wave sagging and hogging moments will be constant. The still water moment can vary according to the loading of the ship and may, therefore, affect the total moment significantly. The aim in any operational loading condition for a ship is to minimise the total moment.

7.4 Structural response

Once the shear force and bending moments have been found, the stresses in the structure can be calculated. Simple beam theory provides the relationship:

$$\frac{\sigma}{z} = \frac{M}{I} \quad \text{or} \quad \sigma = \frac{Mz}{I} = \frac{M}{Z}$$

where σ is the stress at some distance z from the neutral axis of the section, M is the bending moment at the section considered and I is the second moment of area about the neutral axis of the section at x. $Z = I/z$ and is known as the section modulus.

The maximum stresses will occur when z is a maximum, ie at the deck and at the bottom shell plating of the ship. The neutral axis may not be at the mid-depth of the section resulting in two values of z, one for the maximum tensile stress and one for the maximum compressive stress.

The application of this theory has been found to give acceptable results when applied to the complex structure of a ship. When determining the longitudinal stress for a known bending moment the values of I and Z must be established for the cross-section where the maximum bending moment occurs. This will be at, or close to, midships. A cross-section at midships and also where there are openings may be used in the calculations, such as shown in Fig 7.6. Only material which is distributed over a considerable length in the fore and aft direction should be included since this will be effective in resisting the bending moment. This would include all continuous decks, deck longitudinals, side and bottom shell plating, bottom longitudinals, tank top plating with margin plate (where fitted) and the centre girder.

An assumed neutral axis at about 40 to 45% of the depth of section above the baseline is usual. The parallel axis theorem can then be applied to determine values about the actual neutral axis.

Example

The items and cross-sectional areas of structure shown in Fig 7.6 are given in the first column of Table 7.1. The neutral axis is initially assumed to be 4.5m above the base. The depth from deck to bottom shell is 12m. If the maximum hogging bending moment is 595.5MNm determine the stress in the deck and bottom plating.

Only half of the ship is considered in the calculation and the result is then doubled. Only half of the area of the bottom centre girder is used in the table, since this is the only item that is not present on each side. Column 3 is the distance, z, of the centroid of the individual item from the assumed neutral axis.

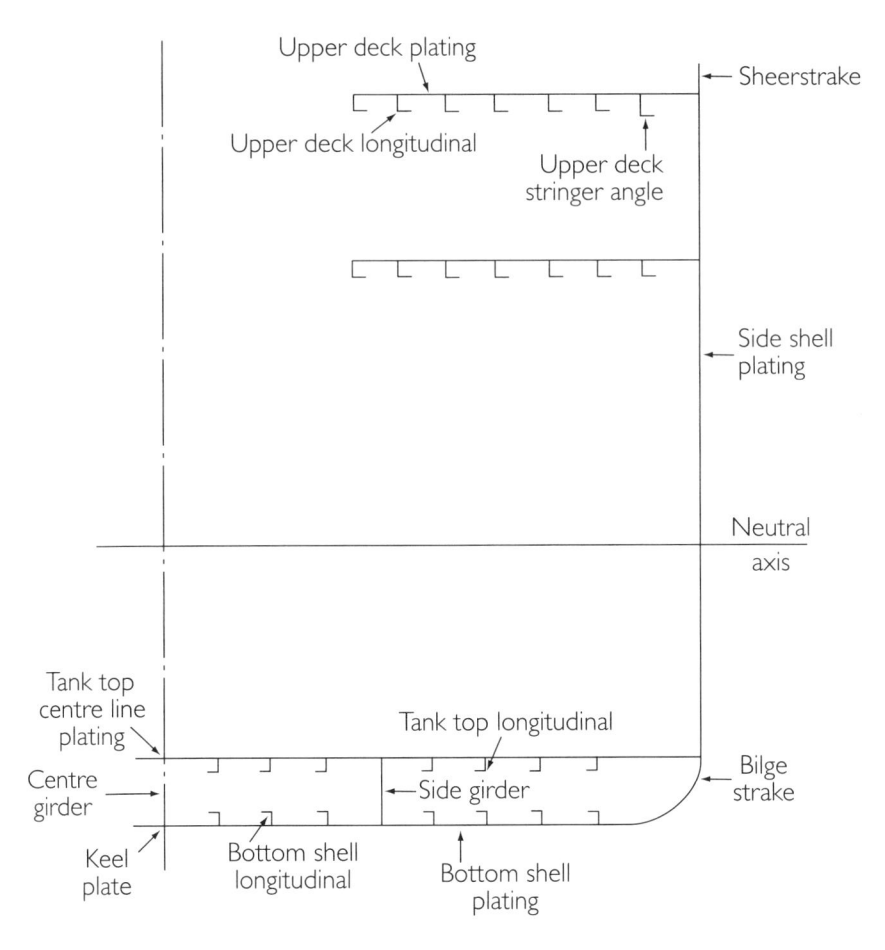

7.6: Midship section of a ship showing longitudinal strength members

The parallel axis theorem is used to determine the moment of inertia of each item about the assumed neutral axis. The parallel axis theorem states:

$I_{na} = I_{xx} - Az^2$

where I_{na} is the moment of inertia about the neutral axis, I_{xx} is the moment of inertia about some axis xx, A is the cross-sectional area of material and z is the distance from the axis, xx, to the neutral axis.

In Table 7.1, column 4 is the product $A \times z$, column 5 the value $A \times z^2$ and column 6 is the moment of inertia about the neutral axis for the item considered. Most of the items can be assumed to be rectangular, so the moment of inertia is $1/12\, Ah^2$, where A is the area of the item, and h is the depth. The moment of inertia of any horizontal material on longitudinals, for example, about its own neutral axis is so small it can be neglected.

Merchant Ship Naval Architecture

1	2	3	4	5	6
Item	Area, A ($m^2 \times 10^{-4}$)	Z (m)	Az ($m^3 \times 10^{-4}$)	Az^2 ($m^4 \times 10^{-4}$)	$Ah^2/12$ ($m^4 \times 10^{-4}$)
Upper deck stringer angle	60	7.42	445	3300	-
Upper deck plating	1210.5	7.73	9350	72 400	-
Upper deck longitudinals	107.2	7.67	820	6300	-
Second deck plating	574.3	4.97	2860	14 200	-
Second deck longitudinals	95.4	4.91	468	2300	-
Sheerstrake	308.2	6.92	2130	14 750	63 (h=1.57)
Side shell	978.4	3.05	2980	9100	3030 (h=6.1)
Total above assumed axis	3334.0		19 053	122 350	3093

Side shell	525.0	1.65	866	1430	470 (h=3.28)
Bilge strake	214.2	3.81	815	3100	46 (h=1.6)
Bottom shell	1240.4	4.51	5590	25 200	
Keel	147.2	4.57	674	3080	
Tank top centre line	90.2	3.42	308	1050	
Tank top plating	935.4	3.42	3200	10 950	
Tank top longitudinal	190.1	3.54	672	2380	
Bottom shell longitudinal	209.5	4.40	924	4060	
Centre girder	78.6	4.00	314	1260	9 (h=1.14)
Side girder	105.0	3.96	415	1640	10 (h=1.09)
Total below assumed axis	3735.6		13 778	54 150	535
Total above assumed axis	3334.0		19 053	122 350	3093
	7069.6		5275	176 500	3628
				3628	
				180 128	

Table 7.1 Calculation for section modulus (Some figures have been rounded up/down for convenience)

The ship's neutral axis has been assumed to be 4.5m above the base resulting in material above and below this axis. The total moment in column 4 is different for material above the assumed axis and material below. The actual neutral axis is above that assumed by a distance, l, where

$$l = \frac{\text{net moment}}{\text{total area}} = \frac{0.5275}{0.7069} = 0.746 \text{m}$$

Neutral axis above base = 4.5 + 0.746 = 5.246m

I about assumed axis (one side) $= 18.0128\text{m}^4$

I about NA (one side) $= 18.0128 - (0.7069.6 \times (0.746)^2)$
$= 17.62\text{m}^4$

I about NA for both sides $= 17.62 \times 2 = 35.24\text{m}^4$

z_1 to upper deck $= 12 - 5.246$ $= 6.754\text{m}$

$\therefore$ Section modulus, $Z_1 = \dfrac{I}{z_1}$ $= \dfrac{35.24}{6.754} = 5.218\text{m}^3$

z_2 to keel $= 5.246\text{m}$

$\therefore$ Section modulus $Z_2 = \dfrac{I}{z_2}$ $= \dfrac{35.24}{5.246} = 6.717\text{m}^3$

Stress in deck $= \dfrac{Mz_1}{I} = \dfrac{595.5 \times 6.754}{35.24} = 114\text{MN}/\text{m}^2$ (tension)

Stress in bottom shell $= \dfrac{Mz_2}{I} = \dfrac{595.5 \times 5.246}{35.24} = 88.7\text{MN}/\text{m}^2$ (compression)

The stress will vary linearly from a maximum tensile stress in the deck (since it is a hogging bending moment) to zero at the neutral axis and then increase linearly to the maximum compressive value in the bottom shell.

7.4.1 Equivalent steel area

Where a ship is constructed of two different materials along a significant proportion of the length, stress calculations are often done using an equivalent steel area. If a simple beam of composite construction is considered then, from beam theory, the stress, $\sigma = E z/R$, where E is the modulus of elasticity, z is the distance from the neutral axis, and R is the radius of curvature. For the beam section to be in equilibrium the net force in the two materials must be zero.

Thus,

$$\sum (\sigma_s A_s + \sigma_a A_a) = 0 \quad \text{or}$$

$$\sum \left(\dfrac{E_s A_s z_s}{R} + \dfrac{E_a A_a z_a}{R} \right) = 0 \quad \text{and}$$

$$\sum \left(A_s z_s + \dfrac{E_a}{E_s} A_a z_a \right) = 0$$

The bending moment will be the product of the force, σA, and its distance, z, from the neutral axis,
thus

$$M = \sum (\sigma_s A_s z_s + \sigma_a A_a z_a)$$
$$= \dfrac{E_s}{R} \sum \left(A_s z^2 + \dfrac{E_a A_a z_a^2}{E_s} \right) = \dfrac{E_s I_e}{R}$$

where I_e is the effective second moment of area of the composite beam.

The composite ship's cross-section can thus be considered made up of material, s, which is normally steel, if an equivalent steel area of material, a, is used. The equivalent steel area is the actual area multiplied by E_a/E_s. The value is effectively unity for different steels and for aluminium alloy/steel would be approximately $1/3$.

7.4.2 Changing the section modulus

The addition of more structural material within the hull, as a result of a design change, may or may not increase the section modulus. Increasing the depth of the section, likewise, may not necessarily result in a reduction in stress.

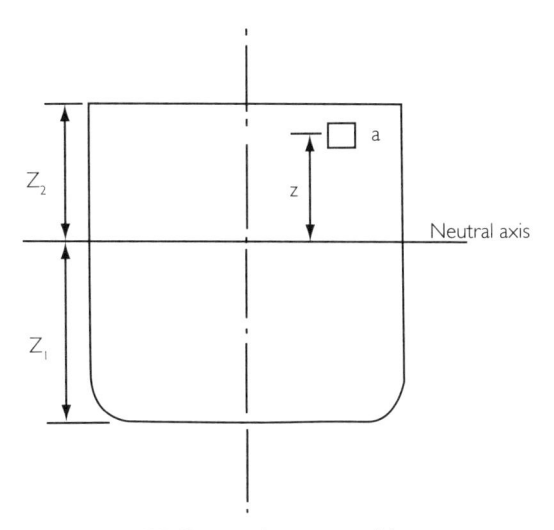

Fig 7.7: Changing the section modulus

An item of structure, area, a, added at distance, z, above the neutral axis, but below the deck will be considered first, see Fig 7.7. If the original cross-section area was A and the radius of gyration, k, then following the addition of the item of area, a, the neutral axis will be raised by an amount, δz, where:

$$\delta z = \frac{az}{(A+a)}$$

The second moment of inertia will become:

$$I + \delta I = I + az^2 - (A+a)\left(\frac{az}{(A+a)}\right)^2$$

thus

$$\delta I = \frac{Aaz^2}{A+a}$$

The stresses for a particular bending moment will not increase if the section modulus is not reduced, ie

$$\frac{I+\delta I}{Z+\delta z} - \frac{I}{z} > 0, \quad \text{or} \quad \frac{\delta I}{I} > \frac{\delta z}{z}$$

For the item of structure considered, δI is positive and $\delta z/z_2$ is negative, thus $\delta I/I$ is always greater than $\delta z/z$ at the deck plating, as long as the material is added within the cross-section. A similar approach can be adopted for the effect on stresses in the keel. To reduce the stresses in the keel, the item must be added at a distance greater than k^2/z_1 from the neutral axis. Where material is added below the neutral axis a similar approach can be followed.

Increasing the depth of the cross-section will not necessarily reduce the maximum stress. If the increase in z is relatively greater than the increase in I, the ship's girder is not strengthened by the addition of the material. If it is assumed that the stress at the highest deck should not exceed that at the top of the main structure, then:

$$\frac{\delta z}{z} = \frac{\delta I}{I}$$

Consider the addition of an area, a, of deck plating added at a height, h, above the main structure of cross-sectional area, A, and moment of inertia, I. Let z_2 be the distance from the original neutral axis to the top of the main structure and let $x = z_2 + h$, see Fig 7.8,

then

shift of neutral axis $= \dfrac{ax}{A+a}$

New I about original axis $= I + ax^2$

New I about new axis

$$= I + ax^2 - (A+a)\left(\frac{az}{A+a}\right)^2$$

$$= I + \frac{Aax^2}{A+a}.$$

For the stress in the highest deck not to exceed that at the top of the main structure,

$$\frac{I}{z_2} = \frac{I_1}{z_1}$$

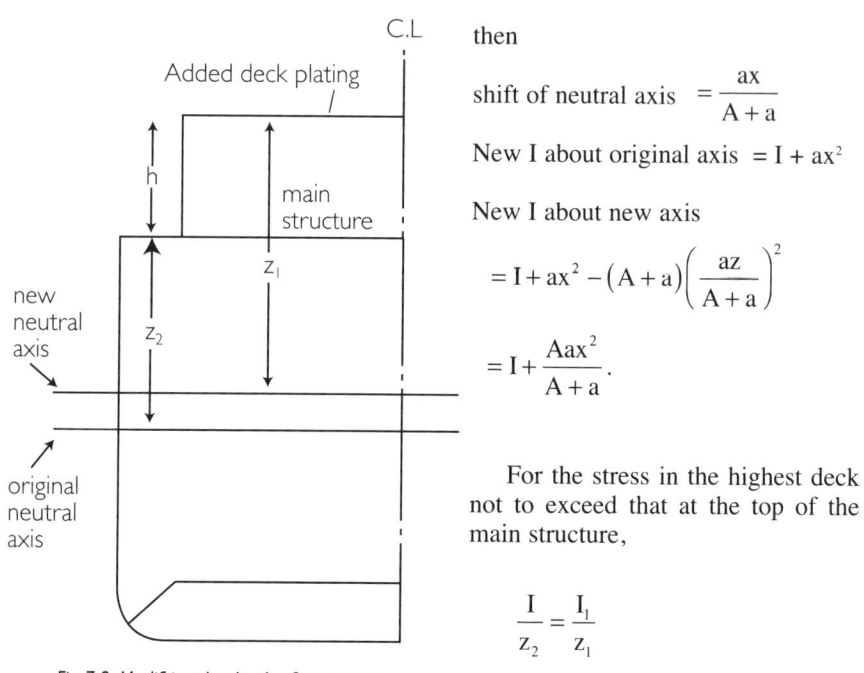

Fig 7.8: Modifying the depth of cross-section

where I_1 is the moment of inertia of the new section and z_1 is the distance from the new neutral axis to the added deck.
Thus

$$\frac{I}{z_2} = \frac{I(A+a) + Aax^2}{Ax}$$

$$a(I + Ax^2) = \frac{IAx}{z_2} - IA = \frac{IA(x - z_2)}{z_2}$$

$$\therefore a = \frac{I\,Ah}{z_2(Ax^2 + I)}$$

Merchant Ship Naval Architecture

Example

A superstructure deck of breadth 9.25m is to be fitted 2.75m above the top of the ship's main structure. Determine the thickness of this deck plating so that the stress in the superstructure deck will not exceed that at the top of the main structure without the superstructure.

The cross-sectional area is 2.26m², second moment of inertia about the neutral axis is 26.8m⁴, the distance from the neutral axis to the top of the main deck is 7.6m.

$$\text{Area of plating, a} = \frac{IAh}{z_2(Ax^2 + I)}$$

Let t be the thickness of the superstructure deck plating in metres, then

$$9.25 \times t = \frac{26.8}{7.6} \times \frac{2.26 \times 2.75}{\{2.26 \times (10.35)^2 + 26.8\}}$$

$$\therefore t = \frac{26.8 \times 2.26 \times 2.75}{268.9 \times 7.6 \times 9.25} = 0.0088 \text{m or } 8.8\text{mm}$$

7.4.3 Stresses when inclined

The standard longitudinal strength calculation assumes that the ship is upright and the bending moment is in the vertical plane. This does not necessarily give the greatest stress in the structure. When a ship is inclined the depth of section is increased and greater stresses will occur at the corners.

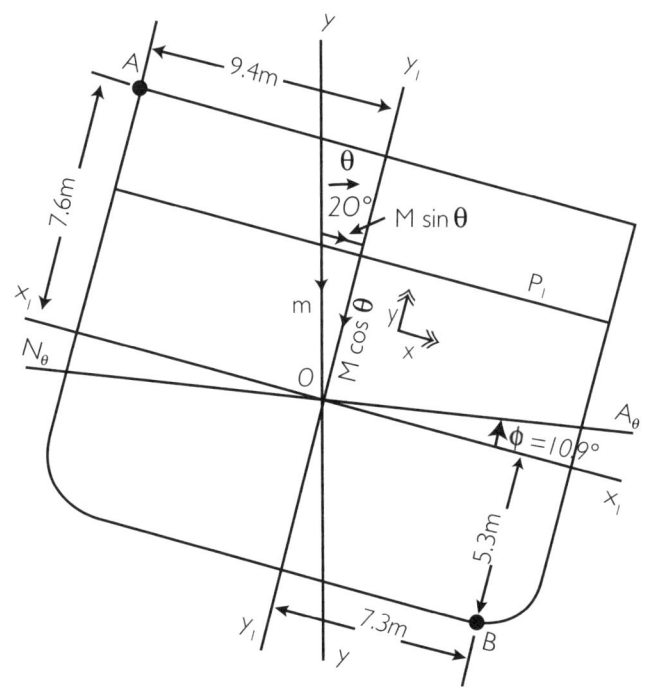

Fig 7.9: Stresses in the inclined position

Consider a ship inclined at some angle θ to the vertical as shown in Fig 7.9. The bending moment will still be in the vertical plane and the problem is one of unsymmetrical bending. Let the bending moment in the vertical plane be m, which can be resolved into two components, M_y in the y plane and M_x in the x plane.

Thus
$$M_y = M \cos\theta \text{ and } M_x = M \sin\theta$$

If x and y are the coordinates of point P in the structure and I_{na} and I_{cl} are the moments of inertia about the horizontal axis in the upright position and about the centreline respectively, then the stress at P, σ_i, will be

$$\text{stress at P, } \sigma_i = \frac{yM\cos\theta}{I_{na}} + \frac{xM\sin\theta}{I_{cl}}$$

At the neutral axis corresponding to the inclined position, $N_\theta A_\theta$ the stress is zero, thus the position of the neutral axis can be found from the condition

$$\frac{y\cos\theta}{I_{na}} + \frac{x\sin\theta}{I_{cl}} = 0$$

thus

$$y = -x\frac{I_{na}}{I_{cl}}\tan\theta$$

This neutral axis position is inclined at an angle, ϕ given by:

$$\tan\phi = \frac{y}{x} = -\frac{I_{na}}{I_{cl}}\tan\theta$$

The neutral axis as inclined, $N_\theta A_\theta$ is a straight line through 0 making an angle ϕ with the neutral axis, $X_1 X_1$ in the vertical condition. In Fig 7.9 the ship is inclined through an angle θ in a clockwise direction, and the angle ϕ is measured from the neutral axis in the counter-clockwise direction.

If $I_{na} = I_{cl}$, then $\tan\phi = -\tan\theta$ and the neutral axis is horizontal. This is unlikely since I_{cl} is normally about twice I_{na} and, therefore, the neutral axis will be inclined at some angle to the horizontal.

The angles at which the greatest and least stresses occur can be found by putting $\delta\sigma_i/\delta\theta = 0$.

$$\frac{\delta\sigma_i}{\delta\theta} = -\frac{yM\sin\theta}{I_{na}} + \frac{xM\cos\theta}{I_{cl}} = 0$$

or $\tan\theta = \dfrac{x}{y}\dfrac{I_{na}}{I_{cl}}$

The greatest stresses can be seen to be associated with the greatest values of x and y which will occur at the corners of the section.

Example

Determine the maximum stresses occurring at points A (9.4, 7.6) and B (7.3, 5.3) when the ship is inclined at 20deg and subject to a hogging bending moment of 520MNm in the upright condition. I_{na} is 34.8m⁴ and I_{cl} is 65.5m⁴.

At an inclination of 20deg,

$$\tan\phi = \frac{y}{x} = -\frac{I_{na}}{I_{cl}}\tan\theta = -\frac{34.8}{65.5} \times 0.3640 = -0.1933$$

$$\phi = -10.9\,\text{deg}$$

For a hogging moment of 520MNm in the upright position, stress at A is σ_A, where

$$\sigma_A = M\left(\frac{y\cos\theta}{I_{na}} + \frac{x\sin\theta}{I_{cl}}\right)$$

$$= 520\left(\frac{7.6 \times 0.9397}{34.8} + \frac{9.4 \times 0.3420}{65.5}\right)$$

$$= 520(0.2052 + 0.0491) = 132\,\text{MN}/\text{m}^2.$$

At point B, the stress will be σ_B, where

$$\sigma_B = 520\left(\frac{5.3 \times 0.9397}{34.8} + \frac{7.3 \times 0.3420}{65.5}\right)$$

$$= 96\,\text{MN}/\text{m}^2.$$

The stresses at the deck and bottom shell in the even-keel situation would be 114 MN/m² and 79MN/m² respectively.

7.5 Local stresses

Stresses have been considered so far in a general sense as applying to the whole structure of the ship. Local stress concentrations can occur at various places throughout the ship for a variety of reasons. These include heavy concentrated loads, discontinuities in the structure, manufacturing processes, operational loading, vibrations and fatigue.

Heavy concentrated loads require adequate support and suitable arrangements to spread the loading throughout structure. Discontinuities in the structure occur, for example, at the ends of superstructures and edges of hatch openings. Plating is usually made thicker and suitably-radiused corners are provided to spread the loads. Stress concentrations may be introduced as a ship is built, due to poor welding or misaligned structural members. Good quality control during the building process will minimise these problems.

The movement of a ship in a seaway can result in forces being generated which are largely of a local nature. When a ship is heaving and pitching, the forward end enters and leaves the water with a slamming effect. This slamming down of the forward region on to the water is known as **pounding**. Additional stiffening material is fitted in the pounding region to reduce the possibility of damage to the structure. The movement of waves along a ship causes fluctuations in water pressure on the plating. This tends to create an in-and-out movement of the shell plating, known as **panting**.

Additional structure is provided at the bows and stern, where this effect is greatest.

Transverse stresses can result from the static pressure of the surrounding water and, while much less than longitudinal stresses, adequate stiffening is still required to prevent distortion of the structure. The rolling of a ship will also result in forces on the structure tending to distort it. This condition is known as *racking* and its effect is greatest when the ship is in the light or ballast condition. The brackets and beam knees joining horizontal and vertical items of structure are used to resist this distortion.

Vibrations set up in a ship due to reciprocating machinery, propellers, etc, can result in the setting up of stresses in the structure. These are cyclic stresses that could result in fatigue failure of local items of structure leading to more general collapse. Balancing of machinery and adequate propeller tip clearances can reduce the effects of vibration to acceptable proportions.

Fatigue failure is of particular concern, since the repeated stressing of an element can cause a surface crack to occur, which may, ultimately, propagate through the structure. Fatigue failures can occur at stress levels lower than yield stress. Also the varying amplitude of stresses occurring in a ship further complicates the determining of a permissible lower level of stress at which fatigue failure will not occur.

7.5.1 Superstructures

Superstructures and deckhouses create major discontinuities in the structure of the ship which forms a girder to resist longitudinal bending. They contribute to the longitudinal strength, but not 100%, since they do not extend the full length of the ship. This presents two problems, first, the need to disperse the localised stresses into the structure and, second, to determine the extent to which a superstructure can contribute to longitudinal strength. The latter problem will now be considered in detail.

A superstructure is fastened to the upper deck along its bottom edge. As the ship hogs the deck plating will arch upwards and extend. The base of the superstructure is also extended, while its top will be compressed. The superstructure is effectively bending in the opposite direction to the ship's hull. If the two are not to separate, forces will be set up. Shear forces will result from the stretching action and normal forces will act to oppose them. The measure of a superstructure's ability to accept these forces and, thus, contribute to the section modulus and resist bending is known as the *superstructure efficiency*, where

$$\text{superstructure efficiency,} \quad \eta_s = \frac{\sigma_0 - \sigma_A}{\sigma_0 - \sigma}$$

where σ_0 is the upper deck stress if no superstructure is fitted, σ_A is the stress calculated in the upper deck and σ is the stress in the upper deck if a fully effective superstructure were fitted. Superstructures can be made more efficient by increasing their length, making them the full width of the ship and maintaining their cross-section constant.

Example

An aluminium superstructure deck 13m wide and 12mm thick is to be added 2.6m above the upper deck on a new ship design. The ship must withstand a sagging bending moment of 450MNm. The stress in the upper deck with the superstructure deck fitted is 55MN/m². What will be the superstructure efficiency?

For the midship section:
Cross-sectional area of longitudinal material   = 2.3m²
Distance from neutral axis to upper deck   = 7.6m
Second moment of area about neutral axis   = 58m⁴
Since this is a composite structure, the second moment of an equivalent steel

section must be found. The stress in the steel sections can then be found and, after the use of the modular ratio, the stress in the aluminium.

New area of material (as effective steel) = 2.3 + (13 x 0.012) × modular ratio

Now modular ratio $= \dfrac{\text{Young's modulus for aluminium}}{\text{Young's modulus for steel}} = \dfrac{67\,000}{208\,000} = 0.322$

Thus:
new area of material (as effective steel) = 2.3 + ((13 x 0.012) × 0.322) = 2.35m².

Movement of neutral axis due to adding deck

$$= \dfrac{0.322 \times (13 \times 0.012)(7.6 + 2.6)}{2.35} = 0.218\text{m}.$$

Second moment of new section about old neutral axis
= 58 + (0.322 (13 x 0.012) (7.6 + 2.6)²) = 63.23m⁴

Second moment of new section about new neutral axis, I
= 63.23 − (2.35(0.218)²) = 63.12m⁴

Distance to new deck from new neutral axis, z
= 7.6 + 2.6 - 0.218 = 9.98m

Stress in new deck (as effective steel)

$$= \dfrac{M}{I_1/z_1} = \dfrac{450}{63.12/9.98} = 71.15\text{MN}/\text{m}^2$$

Stress in new deck (as aluminium alloy) $\sigma_a = 0.322 \times 71.15 = 22.92\text{MN/m}^2$

The superstructure efficiency relates to the effect of the superstructure on the stress in the upper deck of the main hull,

superstructure efficiency, $\eta_s = \dfrac{\sigma_0 - \sigma_A}{\sigma_0 - \sigma}$

The stress in the upper deck with the aluminium superstructure fitted, σ_A, is given as 55MN/m².

With no superstructure fitted the stress, $\sigma_0 = \dfrac{M}{I/z} = \dfrac{450}{58/7.6} = 58.97\text{MN}/\text{m}^2$

For a fully effective superstructure, the stress $\sigma = \dfrac{M}{I_1/z}$

where z is the distance from the new neutral axis to the upper deck

$$\therefore \sigma = \dfrac{450}{63.12/(7.6 - 0.218)} = 52.63\text{MN}/\text{m}^2$$

Superstructure efficiency $= \dfrac{58.97 - 55}{58.97 - 52.63} = 0.63 \text{ or } 63\%$

7.6 Shear stresses

In addition to the bending stress set up in a ship's structure there are also shear stresses. When a ship is moving through waves the shearing forces are at a maximum about one quarter of the length from each end. Large shear forces can also be set up by cargo loads in a ship lying in still water. The maximum shearing force will occur in the region of the neutral axis. Shear forces can result in an increase of bending stresses at the corners of the deck and the round of bilge, and reductions at the centre of the respective structures. They may also increase deflection.

The shearing force at a cross-section will tend to move one part of the ship vertically with respect to the adjacent section. The shear stress set up will be the shear force divided by the area considered. The shear stress at any point, distance y from the neutral axis of a beam, is given by:

$$\text{shear stress} = \frac{FA\acute{y}}{It}$$

where :
F = shear force at the section under consideration
$A\acute{y}$ = the moment of area about the neutral axis above or below the surface under consideration
I = second moment of area of the complete section about the neutral axis
t = total thickness of material resisting shear.

Example

The maximum shearing force in a ship is 44.76MN. The moment of area about the neutral axis is 15.10m³ and the second moment of the section about the neutral axis is 275.8m⁴. The thickness of the side shell plating at the neutral axis is 0.0214m. Calculate the shear stress at the neutral axis.

$$\text{Shear stress} = \frac{FA\acute{y}}{It}$$

$$= \frac{44.76 \times 15.1}{2 \times 0.0214 \times 275.8} = 57.26 \text{MN}/\text{m}^2$$

7.6.1 Deflection of ships

The longitudinal deflection of a ship is mainly due to a change of bending moment, but shear stresses can also contribute. If it is assumed that a ship behaves as a variably loaded and variably supported girder then stresses and deflections can be assessed using simple beam theory.

The quantities, curvature (1/R), slope, and deflection are interconnected in the same way as load, shearing force and bending moment. Although 1/R = M/EI, it is generally more convenient to prepare a curve of M/I and to divide the derived deflection by E.

Thus, the deflection, y, caused by bending is:

$$y = \frac{1}{E}\iint \frac{M}{I}\delta x \delta x,$$

where
y = deflection (when adjusted)
M = bending moment
I = second moment of area
E = modulus of elasticity

For a simply-loaded beam, the value of M is obtained by a simple mathematical expression and, if the beam is of constant cross-section, then I is constant and the integration is straightforward. For a ship, the calculation of the deflection is not so straightforward, since the bending moment cannot be represented by a simple expression and the value of I is not constant over the length of the ship.

The value of I is therefore calculated at a number of positions along the ship's length and a curve of M/I is then plotted. Integration of the curve can be done graphically or in a tabular form by dividing the length of the ship into a number of sections and taking the mean ordinate for each.

The first integration of M/I will give a curve of slope, and the second will give a curve of deflection. The ordinates of this curve are the total deflection occurring between the ends but, since it is usual to measure deflection with reference to a straight line joining the ends, the final deflection curve is obtained by measuring between such a line and the second integral curve and setting up these distances on a horizontal base line.

Example

Estimate the deflection of a ship's hull at amidships relative to the straight line through its extremities at the perpendiculars, when the values of M/I at equidistant stations commencing at the after perpendicular are:

0, 9.126, 13.25, 14.72, 15.31, 14.23, 11.68, 7.65, 0
The ship length is 128m and Young's Modulus, E = 209 000MN/m2
The common interval = 128/8 = 16m.

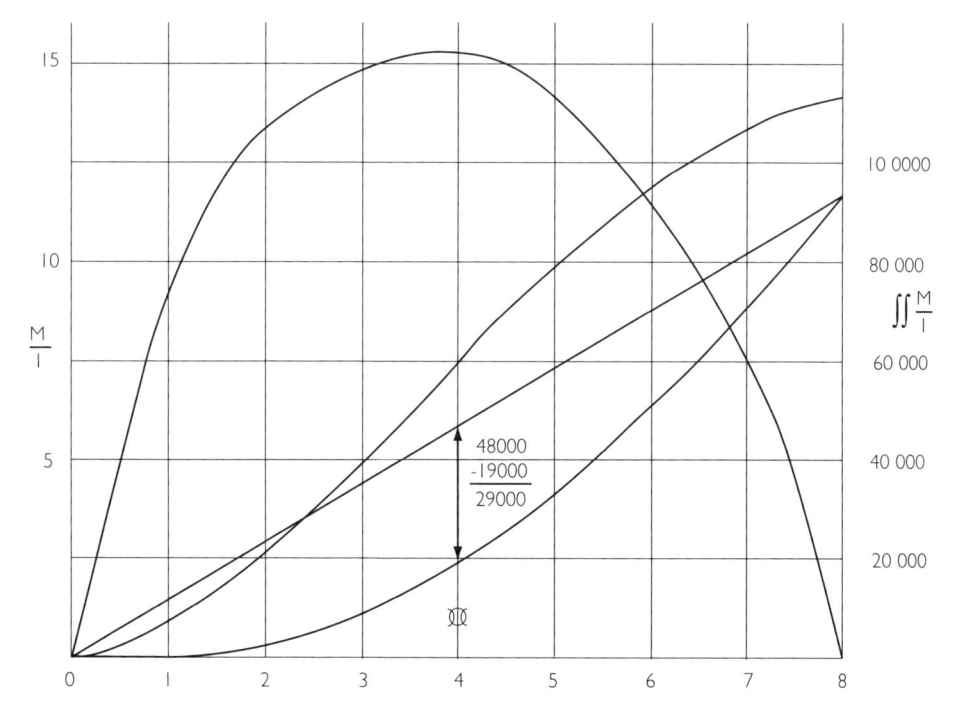

Fig 7.10: Calculation of deflection

The values of M/I are plotted as in Fig 7.10 and mean ordinate values measured and inserted in Table 7.2. The first integral curve can then be drawn and mid-ordinate values determined from it for the second integral curve which is also drawn. The final deflection curve is drawn from zero to the end point of the second integral curve.

Position	Value for first integral	Position	Value for second integral
To station 1	16 × 5.0 = 80	To station 1	16 × 30 = 480
1-2	16 × 11.80 = 188.8	1-2	16 × 170 = 2720
To 2	= 268.8	To 2	= 3200
2-3	16 × 14.2 = 227.2	2-3	16 × 375 = 6000
To 3	= 496.0	To 3	= 9200
3-4	16 × 15.2 = 243.2	3-4	16 × 620 = 9920
To 4	= 739.2	To 4	= 19120
4-5	16 × 15 = 240	4-5	16 × 875 = 14 000
To 5	= 979.2	To 5	33 120
5-6	16 × 13.0 = 208	5-6	16 × 1090 = 17 440
To 6	= 1187.2	To 6	= 50 560
6-7	16 × 9.7 = 155.2	6-7	16 × 1270 = 20 320
To 7	= 1342.4	To 7	= 70 880
7-8	16 × 4.7 = 75.2	7-8	16 × 1380 = 22 080
To 8	= 1417.6	To 8	= 92 960

Table 7.2 Calculation of deflection

The value of the second integral of M/I at midships, measured from the final deflection curve, as shown in Fig 7.10 is 29 000.

Thus deflection, $y = \dfrac{29\,000}{209\,000} = 0.1388 \text{m}$

The deflection resulting from shear is not as readily found. If the shear stress is assumed to be evenly distributed over the vertical elements of the section then:
shear stress = shear force $/A_s$
where A_s is the area of the vertical elements in the section.
For a short length, δx, the shear deflection, δy, will be

$$\delta y = \frac{F\,\delta x}{A_s C},$$

where C is the shear modulus. The value of shear deflection, y, can be found by integration. Shear deflections are generally only 10 to 20% of bending deflections.

7.7 Absolute and classification society stresses

The standard strength calculation gives values of stress that cannot be directly related to those measured on a ship at sea. They have no real meaning in absolute terms. Historical data based on actual ship measurements have established maximum allowable stresses in terms of the length of a ship. Two formulae which have been established and give similar results are:
maximum allowable stress = $77 + 0.25L$ MN/m^2
and
maximum allowable stress = $23(L)^{1/3}$ MN/m^2
where L is in metres.

Values determined from the two formulae can be seen to be in reasonable agreement in Table 7.3.

Length (m)	60	100	140	180	220	260	300
Maximum allowable stress = 77 + 0.25L	92	102	112	122	132	142	152
Maximum allowable stress = 23(L)$^{1/3}$	90	106	119	129	138	146	154

Table 7.3: Values of allowable stress based on two different formulae

If an approximation of the maximum values of bending moment and shearing force on a ship are required early on in the ship design process, this can be done. The maximum bending moment can be related to the product of displacement and length. Thus:

maximum bending moment $\propto \Delta \times L$

or BM max $= \dfrac{\Delta L}{C} \times 9813 \text{Nm}$

where D is the total displacement and C is a coefficient.

The coefficient C depends upon a number of factors, eg load distribution, type of ship and distribution of buoyancy, and thus a general value of C is difficult to obtain. Values of C are generally given for different types of ship at different conditions of load, see Table 7.4.

Ship type	L (metres)	Δ (tonnes)	C (sagging)	C (hogging)
General cargo	130	12 600	35	33
Tanker	160	22 500	40	95
Tanker	220	62 500	40	115
Cargo liner	225	41 000	116	30
Cargo liner	290	64 000	80	30

Table 7.4: Values of constant C for different ship types and loading conditions

The maximum shearing force may be related to displacement and in the standard condition of loading may be taken as:

Maximum shearing force $= \dfrac{\Delta}{10} \times 9813 \text{N}$

Another expression used is:

Maximum shearing force $= \dfrac{C \times BM_{amidships}}{L} \times 9813 \text{N}$

where the coefficient, $C \cong 4$.

The International Association of Classification Societies has established a common standard for longitudinal strength based upon formulae.

The still water loading, bending moment and shear forces are determined by the methods described earlier. The wave induced bending moments and shear forces are then determined from formulae and added.

The wave induced bending moments are calculated as follows:

hogging BM = $0.19 MCL^2 BC_B$ kNm

sagging BM = $-0.11 MCL^2 B(C_B + 0.7)$ kNm

All dimensions are in metres and $C_B \geq 0.6$
The coefficient C is determined as follows:

$C = 10.75 - \left(\dfrac{300 - L}{100}\right)^{1.5}$ for $90 \leq L \leq 300$ m

$= 10.75$ for $300 < L \leq 350$ m

$= 10.75 - \left(\dfrac{L - 350}{150}\right)^{1.5}$ for $350 < L \leq 500$ m.

Distance from stern	0 - 0.4L	0.4L - 0.65L	0.65L - L
M	$\dfrac{2.5x}{L}$	1	$\dfrac{1.0 - (x - 0.65L)}{0.35L}$

Table 7.5 Value of the distribution factor M

M is a distribution factor, which varies according to the position considered along the length of the ship, see Table 7.5.

The wave induced shear forces are calculated as follows:

hogging SF = $0.3 F_1 CLB(C_B + 0.7)$ kN

sagging SF = $-0.3 F_2 CLB(C_B + 0.7)$ kN

The values of F_1 and F_2 vary along the length of the ship as shown in Table 7.6.

Distance from stern/length	0	0.2-0.3	0.4-0.6	0.7-0.85	1
F1	0	0.92F	0.7	1.0	0
F2	0	0.92	0.7	F	0

Table 7.6 Values of F1 and F2

where $F = \dfrac{190 C_B}{110 (C_B + 0.7)}$

At any point between the values in the table the variation is linear. These formulae can be applied to a wide variety of ships but any unusual design aspects that may affect longitudinal strength will have to be reviewed by the classification society.

7.8 Calculating dynamic longitudinal strength

The standard longitudinal strength problem analyses the ship's structure as a beam and considers a worst case situation for a ship poised on a wave equal to the ship's length. The dynamic behaviour of the ship in moving waves and the hydrodynamics of the waves themselves is not considered. As statistical methods of describing the varying forces in waves have been developed, a different approach to the loading on a ship's structure was made possible. Also new methods of analysing the stresses in a ship's structure enabled more accurate assessment of these stresses if the loading was known.

However, the statistical nature of the data used and various hypotheses and assumptions made, means that the results obtained are of a probabilistic nature, rather than deterministic. This probabilistic result is an estimate of the most probable largest bending moment during some specified service lifetime of the ship, which is usually defined as 10^8 wave encounters.

The design value is established from three independent statistical distributions, the short-term distribution in each stationary wave spectrum which when combined with the distribution of these spectra, or the weather distribution, produces the long-term distribution which is considered valid for the ship's lifetime at sea. Crude comparisons can be made between the probabilistic and deterministic methods which indicate the latter overdesigns the ship, in terms of the strength of the structure. A probabilistic analysis is always undertaken for a new ship design and is accepted as the best method to consider the total loading of the ship's structure. The deterministic analysis does have the advantage of producing actual values, which can be used for comparisons between different designs.

The data used in the probabilistic analysis can be obtained from theoretical considerations, model tests and actual measurements of waves and strain levels on ships. The analysis of the variations in wave forces using statistics has been discussed in Chapter 6 and applied to rolling, pitching and heaving motions. This approach considers that calculations of ship behaviour in regular sinusoidal waves can be used to represent the behaviour in actual irregular waves[1]. The dynamic motions of heaving and pitching have an effect on longitudinal strength, in addition to the wave forces and loading on the ship, and the probabilistic approach to dynamic longitudinal strength takes all these forces into account.

7.8.1 Strip theory

The 'strip theory'[2] method considers the forces on a ship due to static buoyancy, waves and accelerations as it moves through the water. The ship is considered to be divided into a number of transverse strips or sections and the various forces acting can be determined, together with allowances for added mass and damping. The effects are first determined for a ship passing through regular waves. The bending moments can then be obtained by totalling the forces and moments acting over the ship's length. Irregular waves can be considered by analysing such seas into regular components and determining the ship's response to each. The total response of the ship in terms of bending moment can thus be found, for any particular sea state.

Response amplitude operators for the bending moments can then be combined with the wave energy spectra to determine the bending moment energy spectra. If more than one type of sea is to be considered, the data would have to be modified to take into account different weather intensities and the probable time a ship would be present in such weather intensities. This would ultimately lead to the probability of a particular maximum bending moment occurring over the assumed lifetime of the ship.

7.8.2 Model tests

Response amplitude operators can be obtained from theoretical considerations, as outlined above, and also by model tests conducted in towing tanks. The forces acting on the models can be measured using balances on hinged models or strain gauges on metal-framed wax models. The model is run in a tank in regular waves of given height and length. The bending moments for any particular model speed can be found and are generally assumed to be directly proportional to the wave height.

The bending moment response amplitude operator can be plotted to a base of frequency of encounter. A non-dimensional coefficient can be created by dividing the response amplitude operator by the square of the ship's length, the breadth, the sea water density and the acceleration due to gravity. This coefficient can then be applied to full size ships, if it is assumed there is no scale effect.

The bending moment coefficient plot can be combined with the wave energy spectra to determine the bending moment energy spectra. If more than one type of sea is to be considered, the data would have to be modified to take into account different weather intensities and the probable time a ship would be present in such weather intensities. This would ultimately lead to the probability of a particular bending moment occurring over the assumed lifetime of the ship.

Example

A ship model is tested in waves with varying frequency of encounter, ω_e. Bending moment response amplitude operators, (M/h) were then found as shown:

Bending moment RAO (M/h) (MN)	0	103	120	106	95	77	64
Frequency of encounter, ω_e	0	0.4	0.8	1.2	1.6	2.0	2.4

A sea spectrum for the ship, which has been corrected for variations of weather intensity and relative time spent in the various weather groups, is:

Wave spectrum ordinate $S(\omega_e)$ (m/s^2)	0	0.106	0.325	0.300	0.145	0.060	0
Frequency of encounter, ω_e (rad/s)	0	0.4	0.8	1.2	1.6	2.0	2.4

The bending moment M is the sum of hogging and sagging moments and the percentage of their occurrence can be considered as 60% hogging and 40% sagging. The ship is assumed to spend 300 days per year at sea for a ship life of 25 years, with an average encounter period of 6s. Calculate the value of bending moment which will only be exceeded once.

Response spectrum ordinate, $E(\omega_e) = S(\omega_e) \times [R(\omega_e)]^2$

The total response of the ship is then found, ie,

$$\text{total response} = \int_0^\infty E(\omega_e) d\omega_e$$

This integration can be performed using Simpson's Rule:

ω_e	$S(\omega_e)$	$R(\omega_e)$	$[R(\omega_e)]^2$	$E(\omega_e)$ $= S(\omega_e)$ $[R(\omega_e)]^2$	Simpson's multiplier	Product for total response
0	0	0	0	0	1	0
0.4	0.106	103	10 609	1124.6	4	4498.4
0.8	0.325	120	14 400	4680.0	2	9360.0
1.2	0.300	106	11 236	3370.8	4	13 483.2
1.6	0.145	95	9025	1308.6	2	2617.2
2.0	0.060	77	5929	355.7	4	1422.8
2.4	0	64	4096	0	1	0
					Σ	31 381.6

Total response $= \dfrac{0.4}{3} \times 31\,381.6 = 4184.2 \text{MN}^2 \text{ m}^2/\text{s}^2$

Total number of stress cycles in the life of the ship

$$= \dfrac{3600 \times 24 \times 300 \times 25}{6} = 1.08 \times 10^8$$

The probability that a particular bending moment, M, will exceed some value, M_j is:

$$Q(M > M_j) = \dfrac{1}{1.08 \times 10^8} = 0.926 \times 10^{-8}$$

The Rayleigh distribution of bending moments is given by

$$Q(M) = \dfrac{2M}{E} \exp\left(\dfrac{-M^2}{E}\right)$$

where Q(M) is the probability that M will have a particular value and E is the area of the spectrum. The probability that M will exceed a value M_j is given by:

$$Q(M > M_j) = \exp\left(\dfrac{-M_j^2}{E}\right)$$

ie, $0.926 \times 10^{-8} = \exp\left(\dfrac{-M_j^2}{E}\right)$

Taking natural logarithms of each side of the equation:

$$-18.50 = \dfrac{-M_j^2}{4184.2}$$

$\therefore M_j = 278 \text{MNm} = M_H + M_S$

where M_H and M_S are hogging and sagging bending moments. Also it was given that:

$$\dfrac{M_H}{M_S} = \dfrac{0.60}{0.40}$$

$\therefore M_H = 0.60 \times 278 = 166.8 \text{MN m}$.

The bending moment which will probably only be exceeded once = 166.8MN m.

7.8.3 Full-scale measurements

The increasing amount of statistical data available from measurements made on ships has enabled a probabilistic approach to the dynamic longitudinal strength problem based on actual data. Statistical strain gauge measurements can provide short term (20 to 30 minutes) strain values, and long-term recording enables measurements over a range of weather intensities. The strain values can be converted to stress values, since stress = strain × modulus of elasticity of the material. Stress values can, in turn, be converted into bending moments, using the section modulus for the ship. Some approximations are inevitable in the calculation of the section modulus, resulting in statistical probability values that are useful for ship design comparison purposes.

For any particular weather intensity the records produced as histograms are a Rayleigh distribution of the form:

$f(x) = 2 \times \exp(-x^2/E)$ where $E = (\Sigma x^2)/N$

The probability of the stress exceeding any value x_i can be found by integrating from x_i to infinity. Thus:

$P_s(x > x_i) = \exp(-x_i^2)/E$

Stress records taken in a number of different weather intensities can be similarly analysed to determine the probability of a stress value being exceeded in any of the weather conditions considered. Weather intensities may be measured using, for example, the Beaufort wind scale of 12 values or a five-value scale covering the same range of sea conditions. The weather intensity within the various weather groups will also vary, usually according to a normal distribution.

The combined probability, Q_s, that the stress resulting from a particular weather group will exceed some value x_i, will result for the product of the normal probability for the weather group and the Rayleigh probability for the stress. Since the major part of the ship's life is spent in moderate weather, if this proportion is also introduced into the combined probability, then the long-term probability that a stress x will exceed some value x_i can be found.

This result can be interpreted in terms of the number of cycles of stress during which the stress x_i may only be exceeded once. An estimate can then be made of the number of stress cycles that a ship will suffer in its lifetime, based upon the strain gauge readings. Thus, if a stress cycle has a length of t seconds then there are 3600/t per hour and 3600 × 24/t per day. The total number of stress cycles in the life of a ship can be estimated assuming the number of days at sea to be 300 and the ship's life to be 25 years. If t = 6 seconds then,

total number of stress cycles = 3600 × 24 × 300 × 25/t.

∴ total number of stress cycles = 3600 × 24 × 300 × 25/6 = 1.08 × 10^8.

This can also be expressed as the probability of occurrence of 10^{-8} that the particular stress would be exceeded in the life of the ship. This stress value could be found from a plot of stress against the total probability outlined earlier.

The probability of occurrence value of 10^{-8} has become accepted as a reasonable value for ship structures. If an acceptable stress value has been determined then the structure must be designed so that its probability of occurrence is 10^{-8}.

7.8.4 A comparison of methods

The standard method or deterministic approach and the probabilistic method of determining the strength of a ship's structure have been outlined above. It is interesting to compare the two methods. The results for the probabilistic approach would produce a graph of stress against number of cycles to show the probability that a particular stress value would only occur once within the estimated lifetime of a ship. If the standard longitudinal strength calculation is performed for a standard wave height of L/20, or any other accepted value, then the corresponding maximum stress values can be found. If this maximum stress value is used to find the probability of occurrence, then the value will be generally found to be a very low probability. The general conclusion is that the standard longitudinal strength calculation creates an over-designed structure.

The standard wave height of L/20 has become too severe a requirement for ships in excess of 150m. While, initially, attempts were made to reduce the standard wave height, the present criterion for longitudinal strength is based on the most probable largest bending moment during a ship's lifetime, usually defined as 10^8 wave encounters.

7.9 Complex bending and torsion

Only longitudinal bending of the ship's structure has been considered so far, but wave actions also bring about horizontal bending and torsion. Containerships with their large, open, hatchways require careful consideration of the wave forces creating this loading on the structure. A detailed study of the hydrodynamic forces on the structure is required to evaluate the loading. The yawing and swaying motions of the ship must also be investigated.

Horizontal bending and torsion will be taking place while vertical bending occurs, so the stresses from this complex bending and torsion will be the sum of both. There will generally be some phase difference between the different forces, which will have to be taken into account in their summation.

Maximum horizontal bending moments and stresses are typically less than half of the vertical counterparts and the difference in phase further reduces their additional effect on, for example deck edge stresses, to about 20%.

7.10 Structural element strength

The ship's structure is made up of numerous elements, many of which are stiffened plates. This stiffened plate, or grillage, can be considered as a length of plating with supporting stiffeners such as offset bulb plates or angle bar. It will be supported at its edges, where it joins on to other plates. When subjected to variations in loading, the plate may alternately be in tension or compression. While considerable stress can be accepted in tension, collapse or buckling can occur at very much lower stresses when in compression.

The grillage can initially be considered as comprised of a stiffener and some associated effective breadth of plating creating a beam or column The critical axial buckling load for an ideal, straight, column is given by:

$$P_{cr} = \frac{\pi^2 EI}{l^2}$$

where l is the effective length of the column, I is the second moment of area of the cross-section and E is the modulus of elasticity of the material. This relationship is known as Euler's formula and assumes the column ends are pin-jointed.

The critical stress can be found by dividing by the cross sectional area, A, of the column. Thus :

$$P_{cr} = \frac{\pi^2 EI}{Al^2} = \frac{\pi^2 E}{(l/k)^2}$$

where k is the radius of gyration of the section.

The ratio l/k is known as the slenderness ratio. The edges of a ship's plating are not pin-jointed, but are supported on all four sides by other plating. This will significantly increase the critical stress. Both longitudinal and transverse stiffened panels are used in a ship's structure and the actual and comparative values of critical stress are therefore of interest.

For a long panel (longitudinally stiffened) the buckling stress can be approximated to:

$$p_{cr} = \frac{\pi^2 E t^2}{3(1-v^2)b^2}$$

where b is the breadth of the panel and v is Poissons ratio.

For a broad (transversely stiffened) panel,

$$p_{cr} = \frac{\pi^2 E t^2}{12(1-v^2)S^2}\left[1+\frac{S^2}{b^2}\right]^2$$

where S is the length of the panel.

The ratio of buckling stresses for the two forms of stiffened panels is,

$$\frac{\text{Longitudinally stiffened } p_{cr}}{\text{Transversely stiffened } p_{cr}} = \frac{4}{[1+S^2/b^2]^2}.$$

If a transversely stiffened panel is assumed to have a breadth of five or six times its length, then the ratio becomes almost four. This increased buckling strength of longitudinally stiffened plating explains its widespread use in a ship's structure in preference to transverse stiffening.

Example

Considering the earlier example of the ship with an aluminium superstructure, determine whether a transverse beam spacing of 730mm will be adequate to resist buckling.

This transversely stiffened new deck can be considered as a broad panel with the buckling stress, p_{cr} given by:

$$p_{cr} = \frac{\pi^2 E t^2}{12(1-v^2)S^2}\left[1+\frac{S^2}{b^2}\right]^2$$

Poisson's ratio will be taken as 0.33

$$\therefore p_{cr} = \frac{\pi^2 \times 67\,000 \times (0.012)^2}{12(1-(0.33)^2)(0.73)^2}\left[1+\frac{(0.73)^2}{(13)^2}\right]^2 = 16.81 \text{ MN/m}^2$$

Since the stress in the aluminium alloy deck, σ_a, is 22.92MN/m² and the buckling stress, p_{cr}, is 16.81MN/m², then this deck would fail by buckling.

The transverse beam spacing would have to be reduced to about 620mm, in order to ensure that failure by buckling would not occur.

$$L = \frac{L_e}{k} \cdot \frac{1}{\pi} \sqrt{\frac{\delta_y}{E}}$$

Where L_e = effective length of column
δ_y = yield stress
k = least radius of gyration at column of cross-section $\sqrt{\frac{I}{A}}$

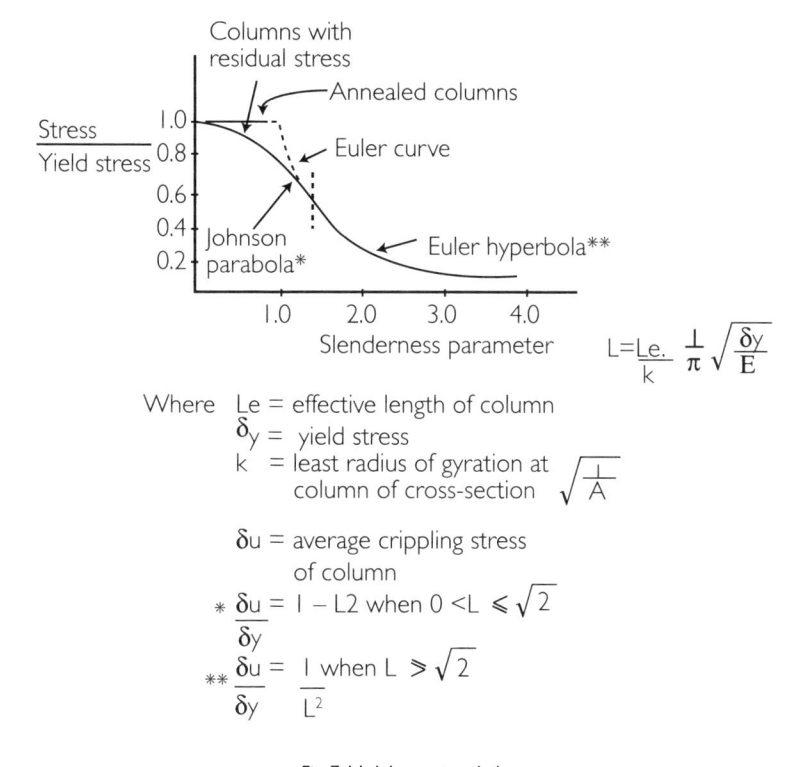

δ_u = average crippling stress of column

* $\dfrac{\delta_u}{\delta_y} = 1 - L^2$ when $0 < L \leqslant \sqrt{2}$

** $\dfrac{\delta_u}{\delta_y} = \dfrac{1}{L^2}$ when $L \geqslant \sqrt{2}$

Fig 7.11: Johnson parabola

The grillages that make up a ship's structure are generally subjected to both lateral and in-plane loading, which leads to higher compressive stresses. Buckling in a practical structure will invariably be inelastic rather than elastic, requiring some modification to the Euler predictions. Inelastic buckling occurs because of curvature of the plates and stiffeners resulting from welding and fabrication deformations. The Johnson Parabola[3] modifies the Euler curve to take into account yielding and buckling, see Fig 7.11.

Considerable research into the strength and stiffness of longitudinally loaded rectangular grillages has resulted in a series of *load shortening curves*[4]. These curves indicate the behaviour of a rectangular plate element both before and after failure. It is important to note that, even after failure, these structural elements can accept some stressing. The curves take into account plate and stiffener characteristics and varying degrees of imperfections related to residual stresses and plate deflection. Load shortening curves enable the design of a more efficient structure, when compared with an approach where yield stress was not exceeded.

7.11 Calculating transverse strength

A ship's structure requires transverse strength to resist stresses and loads created by waves and the motions of the ship. There is also sea water pressure acting on the outside of the hull. Liquid cargoes can create internal loading, due to their movement within tanks and the operations of berthing and drydocking will create occasional loading of the transverse structure.

The analysis of bending and torsional strength assumed no distortion of the ship's cross-section, which means there must be a relationship between transverse strength and both longitudinal and torsional strength. Transverse strength is provided by structural elements such as transverse bulkheads and deep web frames. These elements will assist in transferring loads around the ship's structure, with for example buoyancy pressures on the bottom hull being transferred by the centre girder and longitudinal frames to transverse bulkheads. The bulkheads will transfer some of the load into the side shell. Transverse loads thus become loads that must be considered in longitudinal strength analysis.

Transverse strength needs to be considered for all ships in relation to torsional and racking effects as a result of roll, sway and yaw motions. Loading imposed during shipbuilding and later by drydocking will have to be included in any analysis. Special ship designs with large open areas, such as ro-ro ships, will require careful analysis of their transverse loading. Tankers and ore carriers with loaded centre tanks or holds and empty wing tanks, or vice versa, will also need careful analysis.

Early attempts at analysis of these loads considered a two-dimensional approach, using a 'slice' of the ship's transverse structure. This did not take into account the end fixing conditions of the adjacent structure and its contribution to transverse strength. A strain energy method was developed, which considered the transverse ring of structure as a whole, rather than individual slices. This was, nevertheless, still a two-dimensional approach, which did not consider the effect of adjacent structure in the fore-and-aft direction, such as longitudinal girders.

A three-dimensional approach was subsequently developed which viewed a block of structure between two transverse bulkheads. Detailed calculations, achieved with the use of computers, enabled an analysis of the contribution of longitudinal material to transverse strength to be made. Some elements of structure, eg deep, short, plate girders and large brackets, were not readily analysed. The finite element approach has enabled a more precise analysis of a continuous structure.

7.12 Finite element analysis

Finite element analysis enables the analysis of a ship's structure in terms of primary, secondary and tertiary responses and both longitudinal and transverse strength. The finite element analysis is as follows:
- The structure is divided up into numerous imaginary elements usually of a triangular or rectangular shape, which meet up at nodes.
- A displacement function is used to relate the displacements at any point within the element to that at the node.
- Strains can be determined from the displacement and thus stresses.
- The forces at each node are made equivalent to the boundary forces on the structural element under consideration.
- Displacement so the elements must also be compatible with those of adjacent elements.
- Finally, the whole array of internal forces and applied external forces must be in equilibrium.

Reference should be made to other works[5] for a more detailed description of the technique.

Merchant Ship Naval Architecture

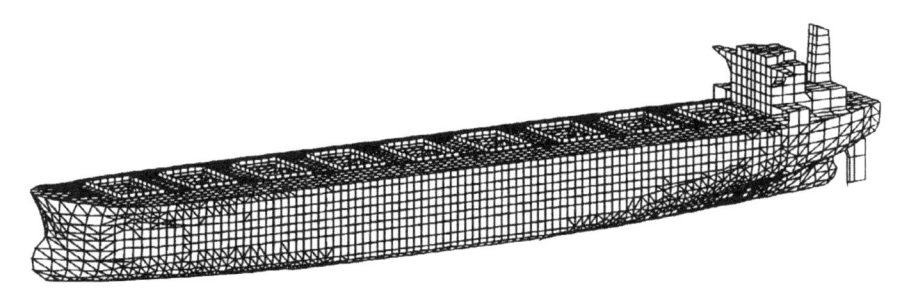

Fig 7.12: Finite element model of a ship (Courtesy of ABS)

The finite element analysis of a complete ship would begin with a coarse mesh of perhaps 1000 nodes and 2500 elements, see Fig 7.12, to obtain an overall response. A finer mesh examination may be done for the whole structure or local, highly-stressed areas.

7.13 Structural failure and safety

A good structural design aims to minimise the probability of failure in the primary, secondary and tertiary structure, which is the main hull, the stiffened plating and the individual plating elements. The loading on the structure will give rise to internal forces that produce stresses, strains and deflections. If the stresses and strains are sufficiently large, the structure will fail. The modes of failure which are likely to affect the primary, secondary or tertiary structure are:

- Plastic collapse when stresses exceed the yield stress in a majority of the structure. Localised yielding may not necessarily lead to collapse, since stresses may be redistributed in the structure, thus sharing out the load.
- Buckling of the structure under compression or shear loads, which generally involves a combination of yielding and buckling in the practical elasto-plastic mode.
- Brittle fracture of a material causing sudden cracking at low values of stress where toughness is lacking.
- Fatigue failure following cyclical loading. The greater the number of load cycles the lower the stress at which such failures will occur.

Appropriate choice of materials will deal with the problem of brittle fracture. The remaining three forms of failure will require some analysis of the loading and the strength of the structure. Where deterministic calculations are made, safety factors are used. It is assumed that these safety factors will cover variations in material properties, assumptions in the calculations, fabrication inaccuracies and operational aspects.

The probabilistic approach seeks to quantify as many of the variations in material properties and assumptions in the calculations, by using statistical data. A probabilistic definition of structural strength is then produced. This enables a probabilistic determination of whether the load on the structure will exceed its strength. Difficulties exist in defining the mathematical nature of the distributions, but load factor values and safety indices have been proposed.

Finally, it remains to assess fatigue strength, which can be done probabilistically, as long as account is taken of the varying nature of cyclical loads on a ship's structure.

7.14 References

1. Denis M, and Pierson W. *On the motions of ships in confused seas,* Trans. SNAME, 1953, 61, 280-332.

2. Korvin-Kroukovsky B. *Theory of Seakeeping*, SNAME, New York, (1961).

3 Faulkner D, Adamchak JC, Snyder GJ, and Vetter MF. *Synthesis of welded grillages to withstand compression and normal loads.* Computers and Structures, 3 (1973).

4. Smith CS, Davidson PC, Chapman JC, and Dowling PJ. *Strength and stiffness of ships' plating under in-plane compression and tension.* Trans. RINA (1988)

5. Zienkiewiczx I. *The Finite Element Method in Engineering Science*, McGraw-Hill, London (1971)

8 Resistance

The accurate prediction of ship resistance is important both to ensure that it is kept to a minimum and to establish the power needed to propel the ship at various speeds. The power required to propel a ship through the water will depend upon the resistance offered by the water and the air, the efficiency of the propulsion system used and the interaction that occurs between them. Propulsion system efficiency and its interaction with the hull will be considered in Chapter 9.

When a body moves through a fluid that is otherwise at rest, friction between them causes a thin layer of fluid to stick to the surface of the body and move with it. At some distance from the body, the fluid remains at rest. The velocity of the fluid is rapid when close to the body, but reduces with increasing distance from it. This region is known as the **boundary layer**. The friction creates a resistance or drag which opposes the movement of the body. Pressure changes also occur around the body leading to a form drag.

Pressure fluctuations around the body as it moves will cause movement of the water surface or waves, which will also have a resistance effect on the body. The total resistance of the body or ship is generally considered to be the sum of the frictional and wavemaking resistances. Since the upper part of the ship is moving through the air, this will also create some additional viscous drag, but this is usually viewed as small in relation to the water-related resistances.

In the assessment of ship resistance it is traditional to group wave-making resistance, form resistance, eddy resistance and air resistance into one force called residuary resistance, R_R. The total resistance, R_T, is then given by;

$$R_T = R_R + R_F$$

where R_F is frictional resistance.

8.1 Ship and model correlation

Models are used to predict full-scale resistance of ships. The criteria which make these predictions possible are based upon the laws of fluid dynamics and some assumptions, which will now be outlined with the aid of dimensional analysis.

If a ship's hull is considered floating in water and the fluid forces with respect to air are ignored, then the resistance of the hull can be considered to be dependent upon the length L, the speed V, the properties of water, eg dynamic viscosity μ, density ρ, bulk modulus K, local absolute vapour pressure p_V, and surface tension ψ. There will also be a need to consider the weight per unit mass of the hull, since there may be deformation of the air/water interface.

Resistance can be written as a functional relationship of these physical variables. Thus,

resistance, $R = f[L, V, \mu, \rho, K, p_V, \psi, g]$

The dimensional formulae of the variables can then be inserted:

$$MLT^{-2} = f\{L\ (LT^{-1})(ML^{-1}T^{-1})(ML^{-3})(ML^{-1}T^{-2})(ML^{-1}T^{-2})(MT^{-2})(LT^{-2})\}$$

The indices of the fundamental dimensions on each side of the equation can be equated resulting in five unknown indices and the following expression for resistance:

$$R = \rho V^2 L^2 f\left[\left(\frac{\mu}{\rho VL}\right)^d, \left(\frac{gL}{V^2}\right)^e, \left(\frac{p}{\rho V^2}\right)^f, \left(\frac{\rho V^2}{K}\right)^g, \left(\frac{\psi}{\rho V^2 L}\right)^h\right]$$

or:

$$R = \rho V^2 L^2 \left[f_1\left(\frac{\mu}{\rho VL}\right), f_2\left(\frac{gL}{V^2}\right), f_3\left(\frac{p}{\rho V^2}\right), f_4\left(\frac{\rho V^2}{K}\right), f_5\left(\frac{\gamma}{\rho V^2 L}\right)\right]$$

These various factors can be viewed as a number of dimensionless groups since $\rho V^2 L^2$ has the dimensions of force or resistance, R. Furthermore, since the groups are dimensionless, it is permissible to rearrange the relationships between them by division, multiplication or raising to a power, which can result in more convenient and familiar expressions and magnitudes for use in experiments. This leads to the following groupings:

$\dfrac{R}{\rho V^2 L^2}$ which is known as the resistance coefficient

$\dfrac{\rho VL}{\mu} = \dfrac{VL}{\nu}$ Reynolds number (Rn)

$\dfrac{V}{\sqrt{gL}}$ Froude number

$\dfrac{P_0 - P_v}{\frac{1}{2}\rho V^2}$ Cavitation number

$\dfrac{\rho V^2}{K}$ Cauchy number

$V\left(\dfrac{\rho L}{\gamma}\right)^{0.5}$ Weber number

In the expression for Reynolds number the kinematic coefficient of viscosity, ν, has been substituted for μ/ρ. In the expression for Cavitation number, the term ρV^2 has been replaced by $\frac{1}{2}\rho V^2$ since this is the dynamic pressure of a fluid stream. Also, if K is the isentropic bulk modulus, the Cauchy number becomes V^2/c^2, where c is the speed of sound in water. The term V/c is known as the **Mach number**.

Mach numbers for fluid flow around a ship are small and their effect on the liquid are insignificant so this term will not be considered further. Weber number is inversely proportional to surface tension forces and, as long as large models moving at moderate to high speeds are used, this term will also be ignored. The motion of the hull is unlikely to induce cavitation, resulting in the removal of this term from considerations of resistance. Cavitation is discussed in a later chapter on propulsion.

The expression for resistance thus reduces to:

$$R = \rho V^2 L^2 \left[f_1\left(\frac{\nu}{VL}\right), f_2\left(\frac{gL}{V^2}\right)\right]$$

The first term can be seen to include the coefficient of viscosity and thus relates to frictional resistance. The second term involves motion under the effects of gravity and relates to wavemaking resistance. With only two terms in the expression for resistance, a major assumption is now made — that they are independent of one another.

Thus resistance can be expressed as:

$$R = \rho V^2 L^2 \left[f_1\left(\frac{v}{VL}\right) + f_2\left(\frac{gL}{V^2}\right) \right]$$

It is now possible to consider the factors necessary for correlation between two ships, or a ship and a model, with respect to their resistance.

If the f_1 term, which relates to frictional resistance, is considered first with subscripts of F for frictional, s for ship and m for model then:

$$R_{Fs} = \rho V^2 L^2 f_1 \left(\frac{v_s}{V_s L_s}\right) \text{ and } R_{Fm} = \rho V^2 L^2 f_1 \left(\frac{v_m}{V_m L_m}\right)$$

thus:

$$\frac{R_{Fs}}{R_{Fm}} = \frac{\rho_s}{\rho_m} \times \frac{V_s^2}{V_m^2} \times \frac{L_s^2}{L_m^2} \times \frac{f_1(v_s/V_s L_s)}{f_1(v_m/V_m L_m)}$$

It can be seen that as long as $v_m/V_m L_m = v_s/V_s L_s$ the terms f_1 will be equal and

$$\frac{R_{Fs}}{R_{Fm}} = \frac{\rho_s}{\rho_m} \times \frac{V_s^2}{V_m^2} \times \frac{L_s^2}{L_m^2}$$

The condition $v_m/V_m L_m = v_s/V_s L_s$ is known as Rayleigh's Law and the quantity v/VL is Reynolds number. Thus the frictional resistance of a ship and model will be the same when $v_m/V_m L_m = v_s/V_s L_s$. If it is assumed that ρ and v are the same for each, then the general condition to be satisfied by a ship and model which are operating in the same liquid is $V_s L_s = V_m L_m$.

If the f_2 term, which relates to wavemaking resistance, is now considered using the subscripts W for wavemaking, s for ship and m for model then:

$$R_{Ws} = \rho V^2 L^2 f_2 \left(\frac{gL_s}{V_s^2}\right) \text{ and } R_{Wm} = \rho V^2 L^2 f_2 \left(\frac{gL_m}{V_m^2}\right)$$

It can be assumed that g is the same for the ship and model, thus:

$$\frac{R_{Ws}}{R_{Wm}} = \frac{\rho_s}{\rho_m} \times \frac{V_s^2}{V_m^2} \times \frac{L_s^2}{L_m^2} \times \frac{f_2(gL_s/V_s^2)}{f_2(gL_m/V_m^2)}$$

It can be seen that as long as $gL_m/V_m^2 = gL_s/V_s^2$ the terms f_2 will be equal and

$$\frac{R_{Ws}}{R_{Wm}} = \frac{\rho_s}{\rho_m} \times \frac{V_s^2}{V_m^2} \times \frac{L_s^2}{L_m^2}$$

The conditional relationship $gL_m/V_m^2 = gL_s/V_s^2$ can be expressed as $V_m^2/V_s^2 = L_m/L_s$ and thus;

$$\frac{R_{Ws}}{R_{Wm}} = \frac{\rho_s}{\rho_m} \times \frac{L_s}{L_m} \times \frac{L_s^2}{L_m^2} = \frac{\rho_s}{\rho_m} \times \frac{L_s^3}{L_m^3}$$

The displacement of a ship is proportional to ρL^3 and thus:

$$\frac{R_{Ws}}{R_{Wm}} = \frac{\text{Displacement}_s}{\text{Displacement}_m}$$

Thus the wavemaking resistance of a ship and model will be the same when $gL_m/V_m^2 = gL_s/V_s^2$ or $V_m/V_s = \sqrt{L_m/L_s}$. This relationship is known as **Froude's Law of Comparison** and the quantity $V/\sqrt{gL}$ is the Froude number. The term **corresponding speeds** is used to indicate that a ship and a model are moving according to the Froude law.

This consideration of frictional- and wavemaking resistance results in two conditions which must be satisfied for ship and model, or ship to ship, correlation, namely:

$V_m L_m = V_s L_s$ and $V_m/\sqrt{L_m} = V_s/\sqrt{L_s}$.

It is only possible to satisfy both conditions simultaneously if the ship and model are the same size. It is usual with model testing to satisfy Froude's Law of Comparison, ie $V_m/\sqrt{L_m} = V_s/\sqrt{L_s}$ where the speeds of the ship and model are in the ratio of the square root of their length ratio. Thus a $1/25$ scale model would be run at $1/5$ of the speed of the full size ship in order to determine the frictional resistance. The actual procedure for model testing will be described in a later section, after a more detailed examination of the components of ship resistance.

8.2 Resistance components

The total resistance of a ship is usually viewed as the sum of frictional and wavemaking resistance. Frictional resistance is an assessment of the effects of the hull friction, and wavemaking resistance considers the wave systems created by the ship.

There will also be some resistance resulting from the ship's form, the creation of eddy currents, appendages on the hull and the air surrounding the upper parts of the hull.

8.2.1 Frictional resistance

Frictional resistance is sometimes called viscous resistance because of its dependence upon viscosity. The variation in velocity of the liquid in the boundary layer is due to shear forces set up in adjacent layers. Some mass of fluid is dragged along with the ship's hull due to viscous action. Flow around the immersed hull occurs in a series of streamlines but as it approaches the stern, the flow separates. Sudden changes in curvature at appendages, such as the stern frame, and the viscosity of the fluid are responsible for this flow separation and the formation of eddies in the fluid. The energy remaining in this fluid and lost from the ship is viewed as an eddy-making resistance.

When a flat smooth surface such as a plate moves through a viscous fluid, the fluid near it is carried along with it. Some distance out from the surface the fluid is not affected by the movement of the plate and remains at rest. The fluid set in motion between the surface of the plate and the undisturbed outer liquid is called the

Merchant Ship Naval Architecture

boundary layer or frictional wake. The movement of water inside the wake may follow two patterns and it is assumed in both cases that the actual water next to the surface has no motion relative to the surface, but is carried along with it. Firstly, if between this and the outer edge of the wake, where the fluid is at rest, the water moves in a series of layers without mixing, the flow is said to be laminar. Secondly, if the flow is such that eddying occurs in the wake, with resulting mixing of the layers, the flow is said to be turbulent.

The flow that will be set up in given circumstances depends upon the relative importance of inertia and viscous forces. The former favours turbulent flow, the latter laminar flow. It can be shown that the ratio of these two forces is represented by the parameter VL/ν, where V is the speed of the surface relative to still water, L is the length of surface in the direction of motion and ν is the coefficient of kinematic viscosity of the fluid. This parameter is called the Reynolds number (R_n) of the motion, thus:

$$R_n = \frac{VL}{\nu}$$

Reynolds number is used as a means of distinguishing between the two distinct types of fluid flow — laminar and turbulent. Laminar flow occurs at low Reynolds number when the fluid particles follow streamlined paths. Turbulent flow occurs where the steady-flow pattern breaks down. Reynolds' work involved liquid flow in tubes and established that different laws of resistance apply to the two types of flow. A critical Reynolds number determines the point at which laminar flow begins to change to turbulent flow.

Reynolds' findings are applicable to the resistance of ships although not in a quantitative sense. The laws of resistance for laminar and turbulent flow are different and tests on the resistance of flat plates have borne this out. Turbulent flow at high Reynolds numbers results in higher resistance than laminar flow as Reynolds predicted. The critical Reynolds number at which laminar flow breaks down is determined by the relationship:

$$\text{Critical Reynolds number} = \frac{VL}{\nu}$$

where V is fluid velocity, L, is the distance from the leading edge of a flat plate and ν is the kinematic viscosity of the liquid.

Prior to the point L on the flat plate the flow is laminar, at L a period of transition occurs and after the transition region fully turbulent flow is developed. These findings are important when conducting model tests in order to ensure that models are run in turbulent flow and not in laminar or transition regions. To prevent the formation of laminar flow at the bows of ship models, some form of turbulence stimulation, such as lengths of wire or lines of studs, are used on the hull surface.

8.2.2 Frictional resistance experiments

William Froude carried out the first investigations into frictional resistance in the 1870s by towing submerged planks in what was the first-ever towing tank, at Torquay in England. Different sizes of planks with various surface roughness were used. Froude concluded that frictional resistance depended upon:
- The area of the surface.
- The type of surface.
- The length of the surface.
- The density of the water.

- The nth power of speed.
  Froude found that frictional resistance could be expressed by the formula:

$$R_F = fSV^n$$

where R_F = frictional coefficient in Newtons,
f = a coefficient depending on the length of the surface,
S = wetted surface area of ship in square metres,
V = speed of ship in metres per second.

Froude's son, RE Froude, standardised the index for speed at 1.825 giving the formula

$$R_F = fSV^{1.825}$$

The values of f in sea water for use with RE Froude's formula are given in Table 8.1.

Length (m)	f	Length (m)	f	Length (m)	f
2.0	1.966	11	1.589	40	1.464
2.5	1.913	12	1.577	45	1.459
3.0	1.867	13	1.566	50	1.454
3.5	1.826	14	1.556	60	1.447
4.0	1.791	15	1.547	70	1.441
4.5	1.761	16	1.539	80	1.437
5.0	1.736	17	1.532	90	1.432
5.5	1.715	18	1.526	100	1.428
6.0	1.696	19	1.520	120	1.421
6.5	1.681	20	1.515	140	1.415
7.0	1.667	22	1.506	160	1.410
7.5	1.654	24	1.499	180	1.404
8.0	1.643	26	1.492	200	1.399
8.5	1.632	28	1.487	250	1.389
9.0	1.622	30	1.482	300	1.386
9.5	1.613	35	1.472	350	1.376
10.0	1.604				

Table 8.1: RE Froude's skin friction constants

8.2.3 ITTC ship-model correlation line

Lord Rayleigh demonstrated a connection between Froude number and Reynolds number, and subsequently a number of investigators assumed that Reynolds number should form the basis of friction calculations rather than the Froude empirical formulae.

The Dimensional Analysis approach examined earlier produced an expression for frictional resistance where

$$R_F = f(\tfrac{1}{2}(SV^2))$$

which can be expressed as:

$$C_F = \frac{R_F}{\tfrac{1}{2}\rho SV^2} = f(R_n)$$

where C_F is a frictional resistance coefficient. Schoenherr examined a large amount of experimental data and produced a formula:

$$\frac{0.242}{C_F^{\frac{1}{2}}} = \log 10(R_n C_F)$$

The plotting of C_F against Reynolds number creates the Schoenherr line, which applies to the frictional resistance of a smooth hull.

The relationship between the frictional resistance coefficient and Reynolds number was standardised by the International Towing Tank Conference (ITTC) in Madrid in 1957 as;

$$C_F = \frac{R_F}{\frac{1}{2}\rho S V^2} = \frac{0.075}{[\log R_n - 2]^2}$$

Values of C_F for the ITTC formula are given in Table 8.2.

Reynolds number, R_n	$C_F \times 10^3$ for $R_n \times 10^6$	$C_F \times 10^3$ for $R_n \times 10^7$	$C_F \times 10^3$ for $R_n \times 10^8$	$C_F \times 10^3$ for $R_n \times 10^9$
1.00	4.688	3.000	2.083	1.531
2.00	4.054	2.669	1.889	1.407
3.00	3.742	2.500	1.788	1.342
4.00	3.541	2.390	1.721	1.298
5.00	3.397	2.309	1.671	1.265
6.00	3.285	2.246	1.633	1.240
7.00	3.195	2.195	1.601	1.219
8.00	3.120	2.152	1.574	1.201
9.00	3.056	2.115	1.551	1.185

Table 8.2: Values of CF for the ITTC formula

When plotted as C_F against Reynolds number, a model-ship correlation line is produced, see Fig 8.1. Since this data is based upon experiments involving smooth flat plates, the results obtained will require some addition for roughness of the full-size ship. The 1978 ITTC Performance Prediction Method proposes the following increase in frictional resistance, δC_F, for roughness,

$$\delta C_F = \left[105\left(\frac{k_s}{L}\right) - 0.64\right] \times 10^{-3}$$

where k_s is the hull roughness and L is the waterline length of the ship. For a new ship, k_s is typically 80-150 μm giving a δC_F in the region of 0.0004.

Further research resulted in the view that C_F obtained from the friction line did not correspond exactly to the model or ship hull, but required the provision of a model-ship correlation allowance. A coefficient of form resistance was introduced to allow for the three-dimensional flow over a hull shape and took into account pressure resistance arising from the curvature of the hull. A viscous resistance value was thus established from the combination of frictional and form resistance effects.

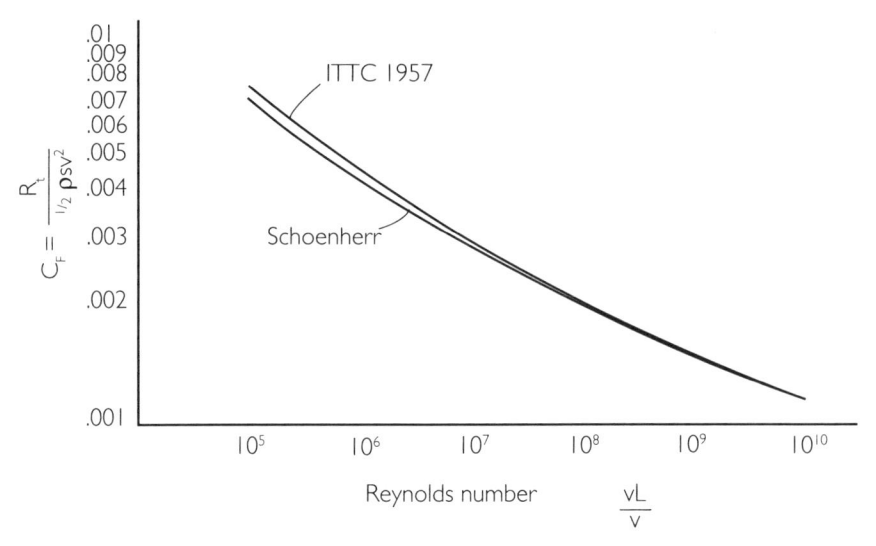

Fig 8.1: Schoenherr and ITTC 1957 model-ship correlation lines

Thus the viscous resistance coefficient C_v can be expressed as

$$C_V = (1 + k) C_F$$

where k is a form factor determined from model tests and assumed independent of speed and scale.

Example

A ship of length 134m and wetted surface area of 3170m² operates at 16kts (1kt = 0.5144m/s). Determine the value of frictional resistance, using the ITTC formula, if kinematic viscosity, v, is 1.19 x 10⁻⁶ and also the value of the Froude number, F_n.

$$\text{Reynolds number, } R_n = \frac{VL}{v} = \frac{16 \times 0.5144 \times 134 \times 10^6}{1.19} = 9.25 \times 10^8$$

$$C_F = \frac{0.075}{[\log R_n - 2]^2} = \frac{0.075}{[6.966]^2} = 0.00154$$

$$R_F = C_F \times \tfrac{1}{2}\rho S V^2 = 0.00154 \times \tfrac{1}{2} \times 1000 \times 3170 \times (8.23)^2$$

$$= 165\,500\text{N}$$

$$\text{Froude number } F_n = 0.3193 \frac{V}{\sqrt{L}} = 0.3193 \times \frac{8.23}{11.57} = 0.227$$

Wetted surface area

Any calculation of frictional resistance will require an estimation of the wetted surface area of the hull. An accurate determination was not readily possible when much of this experimental work was done. Use can be made of Simpson's Rule to integrate the girths of the ship, but even this does not take into account the slope of the sections at the ends.

Approximate formulae have been developed:

$S = L(C_B B + 1.7\,T)$, Denny,

where L is the length, C_B is block coefficient, B is breadth and T is draught;

$S = C\sqrt{\Delta L}$, Taylor,

where C is a coefficient with a value generally about 2.58 and Δ is displacement, in tonnes.

Modern computer programs will allow more accurate estimation of wetted surface area.

8.2.4 Wavemaking resistance

Varying pressures on the hull of a ship as it moves through water result in the formation of waves. The energy required to produce these waves is taken from the moving ship and results in a wavemaking resistance, which is dependent upon the speed and form of the ship's hull.

Three types of wave are generally formed when a ship moves through still water, namely bow and stern divergent and transverse waves. The wave crests of transverse waves move progressively towards the after end and away from the ship. The crests of diverging waves 'fan out' away from the hull.

As the bow pushes through the water it creates a pressure field and initially generates a crest at the bow. A suction field is created at the stern and a trough is initially generated. As the bow waves move aft they interact with the formation of the stern waves resulting in interference effects which may combine to increase their effect or oppose and thus cancel it out. This results in a series of 'humps' and 'hollows' in the curve of wavemaking resistance against Froude number ($V/\sqrt{gL}$), see Fig 8.2.

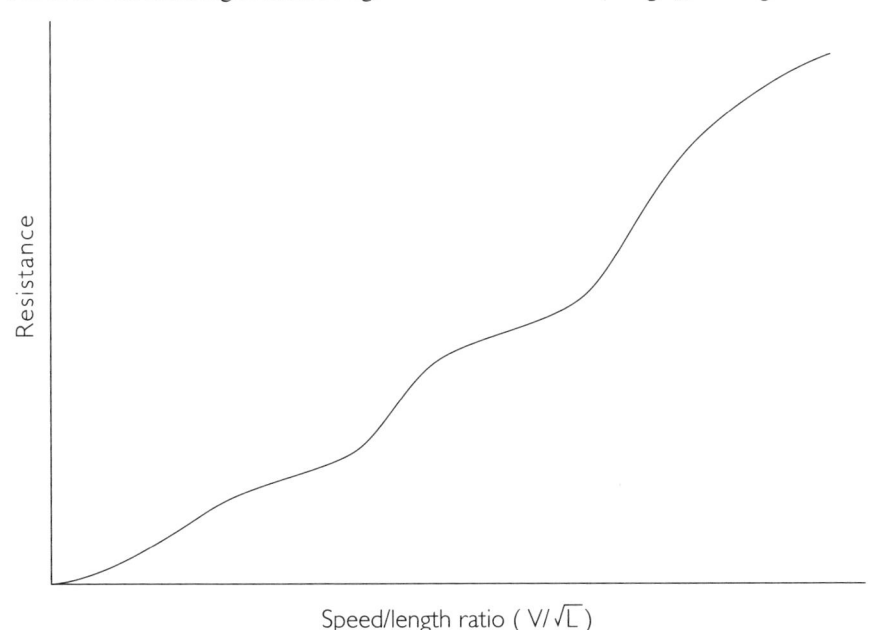

Fig 8.2: Humps and hollows in the wavemaking resistance curve

A more detailed investigation of this effect enables estimation of the Froude number at which the humps and hollows occur for a particular ship design. Ideally a ship should be in a hollow when at its operational speed.

In this examination of wavemaking resistance, the drag effects have been considered solely due to pressure variations creating waves. There is, however, a viscous effect at the boundary layer which would be more evident on a model and needs to be taken into account when scaling-up wavemaking resistance to the full-size ship.

8.2.5 Other resistance effects
In addition to the major resistance elements there are also eddy making, appendage and air resistance to consider, all of which are collectively included with wavemaking resistance to become residuary resistance.

The water particles moving past the hull in streamlines cannot always exactly follow the ship's form, and may break away. The pressure acting on the stern is reduced so that there is a resultant force opposing forward motion. This may be referred to as form drag or resistance. When the streamline flow breaks down, a disturbed volume of water is formed in which water particles revolve in eddies. The energy lost in these eddies can be viewed as a further increase in resistance.

There are various appendages on a ship's hull, such as rudders, shaft brackets, bosses, bilge keels and, sometimes, stabilisers. Since they are shorter than the hull these appendages will run at a lower Reynolds number and their resistance would have to be separately determined. Normally, models are run as naked hulls and then some addition, in the order of about 10%, is added to take account of appendages that would be on the ship's hull.

Air is a fluid and as such will resist the passage of the exposed portions of the ship through it. This resistance has both frictional and eddy-making components and is proportional to the projected area of the above water hull and superstructure and the square of the relative wind velocity. At the full speed of the ship in conditions of no wind, the air resistance is about two to four per cent of the total water-related resistance. In severe weather the air resistance can contribute appreciably to slowing down the ship.

Wind-generated waves can also increase resistance and will also affect ship motions. Reductions in ship speed are normally made to offset the motion problems that are the main consideration.

A propeller when generating thrust will alter the pressure fields at the after end of the ship causing an increase in resistance, which is usually called the *augment of resistance*. It is usually considered as a reduction in propeller thrust and will be considered in more detail in Chapter 9 which deals with propulsion.

8.3 Resistance calculation
William Froude made the first towing tank experiments on models of different shape and form from which it became possible to predict full-scale behaviour. He put forward the idea of dividing the total resistance into the residuary resistance and the frictional resistance of an equivalent flat plate. By observing the wave patterns created by geometrically similar forms at different speeds, Froude found that the patterns appeared to be geometrically identical when the models were at speeds proportional to the square root of their lengths. This speed is known as the **corresponding speed** and led to Froude's Law of Comparison which can be stated as:

If two geometrically-similar forms, ie two ships or a ship and its model, are run at speeds proportional to the square root of their lengths, ie corresponding speeds, then their residuary resistances per unit of displacement will be the same.

This important law makes it possible to estimate the residuary resistance of a ship from that of a model or from a ship of different size, but the same form.

Merchant Ship Naval Architecture

Froude constructed a testing tank from a large trench, cut along a hillside near Torquay. Modern tanks are similar to that first tank, being typically of rectangular cross-section and spanned by a carriage which tows the model along the tank. In a typical run, the carriage is accelerated to the required speed, resistance records are made by a dynamometer and records of trim of the model are made during a period of constant speed and then the carriage is decelerated. Models are generally about 4m to 9m long and ballasted to achieve the correct draught and trim.

The model testing procedure to determine the total resistance, R_T, of a ship is as follows:
1. Measure the total resistance R_{Tm} of a geometrically similar model at its corresponding speed.
2. Estimate the frictional resistance R_{Fm} of the model. Correct this value for water density.
3. Determine the value of the residuary resistance R_{Rm} of the model from the expression $R_{Rm} = R_{Tm} - R_{Fm}$.
4. Using Froude's Law of Comparison, determine the value of R_{Rs} for the ship in sea water. $R_{Rs} = R_{Rm}(L_s/L_m)^3$.
5. Estimate the value of R_{Fs} for the ship.
6. Determine the value of R_{Ts} for the ship using $R_{Ts} = R_{Fs} + R_{Rs}$.

The total resistance thus determined represents the resistance of the smooth or naked hull and does not include appendages such as bossings. If appropriate, an allowance can also be made for air resistance.

Froude developed a constant notation or circular notation system for presenting his results and calculating resistance. The basic unit length was derived from the cube root of the volume of displacement of the ship. Various constants were than developed and described by a letter surrounded by a circle, hence the circular notation description of this method. Froude's law, which considers residuary resistance proportional to displacement for geometrically similar ships run at corresponding speeds, produced a resistance constant, © (circular c), which was proportional to

$$\frac{\text{Resistance}}{\text{Displacement}}.$$

Froude then considered © to be made up of components of residuary and frictional resistance and, with corrections made for differences in friction on the model and the ship, an estimate of ship resistance could be made. Froude's research work was done in Imperial units, and tables are in these units.

8.4 Correlating ship and model results

The foregoing procedure can be related to the ship-model correlation line to demonstrate the extrapolation taking place. Frictional resistance will be measured up to the line and residuary resistance from the friction line up to the curve representing total resistance, see Fig 8.3. The actual position of the friction line is not critical, since it will not affect the total resistance. Some error in the calculation of frictional resistance is acceptable, where it is the same for both ship and model. The slope of the friction line, however, is important, since it relates the ship and the model.

Example
A model 5.2m in length has a total resistance of 67N when towed at 3.5kts. The wetted surface area of the model is 4.25m². Determine the total resistance of a ship of similar form and length 131m, when operating at the corresponding speed. The kinematic viscosity of fresh water is $1.139 \times 10^{-6} m^2/s$ and sea water is $1.188 \times 10^{-6} m^2/s$

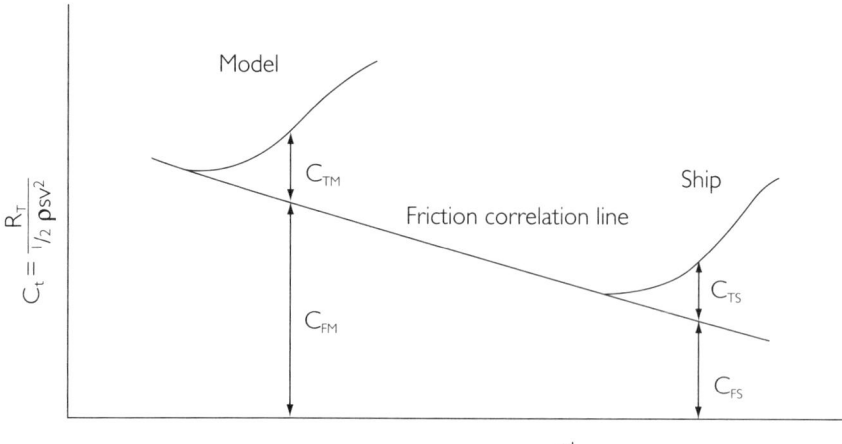

Fig 8.3: Model to ship resistance correlation

Froude coefficient for the model $C_{Fm} = 1.728$ (seawater) and for the ship $C_{Fs} = 1.418$ (sea water).

Model speed = 3.5 knots = 3.5 × 0.5144 m/s = 1.8 m/s

Model frictional resistance, $R_{Fm} = fSV^{1.825} = 1.728 \times 4.25 \times 1.8^{1.825} \times 1/1.025 = 21\text{N}$

Model residuary resistance, $R_{Rm} = 67 - 21 = 46\text{N}$

Corresponding speed of ship = $3.5\sqrt{(131/5.20)} = 17.6\text{kts} = 17.6 \times 0.5144\text{m/s} = 9.05\text{m/s}$

Wetted surface area of ship = $4.25 \times \left(\dfrac{131}{5.2}\right)^2 = 2697.3\text{m}^2$

Ship frictional resistance, $R_{Fs} = fSV^{1.825} = 1.418 \times 2697.3 \times 9.05^{1.825} = 213\,030\text{N}$

Ship residuary resistance, $R_{Rs} = 46 \times \left(\dfrac{131}{5.2}\right)^3 = 827\,398\text{N}$

Ship total resistance, $R_{Ts} = 213\,030 + 827\,398 = 1\,040\,428\text{N} = 1040.4\text{kN}$

ITTC ship-model correlation line:

Frictional resistance coefficient, $C_F = \dfrac{R_F}{\tfrac{1}{2}\rho SV^2} = \dfrac{0.075}{[\log R_n - 2]^2}$

Model total resistance coefficient, $C_{Tm} = \dfrac{\text{model total resistance } R_{Tm}}{\tfrac{1}{2}\rho SV^2}$

$$= \dfrac{67 \times 2}{1000 \times 4.25 \times 1.8^2}$$

$$= 9.73 \times 10^{-3}$$

Merchant Ship Naval Architecture

Reynolds number for model, $R_{nm} = \dfrac{VL}{v} = \dfrac{1.8 \times 5.2 \times 10^6}{1.139}$

$= 8.22 \times 10^6$

Model coefficient of frictional resistance, $C_{Fm} = \dfrac{0.075}{[\log R_n - 2]^2}$

$= \dfrac{0.075}{[6.915 - 2]^2} = 3.10 \times 10^{-3}$

Model residuary resistance coefficient, $C_{Rm} = C_{Tm} - C_{Fm}$
$= [9.73 - 3.10] \times 10^{-3}$
$= 6.63 \times 10^{-3}$

Reynolds number for ship, $R_{ns} = \dfrac{VL}{v} = \dfrac{9.05 \times 131 \times 10^6}{1.188}$

$= 9.9 \times 10^8$

Frictional resistance coefficient for ship, $C_{Fs} = \dfrac{0.075}{[\log R_n - 2]^2}$

$= \dfrac{0.075}{[8.999 - 2]^2}$

$= 1.53 \times 10^{-3}$

Coefficient of total resistance for ship $C_{Ts} = C_F + C_R$ (where C_R is C_{Rm})
$= [1.53 + 6.63]10^{-3}$
$= 8.16 \times 10^{-3}$.

If an allowance of 0.0004 is made for roughness, then C_{Ts} becomes 8.56×10^{-3}

Ship total resistance $= C_{Ts} \times \frac{1}{2}\rho S V^2 = 8.56 \times 10^{-3} \times \frac{1}{2} \times 1025 \times 2697 \times 9.05^2$
$= 969\,049\,N.$

8.5 Standard series data

A natural progression from the testing of individual models is to test some standard form and then vary different parameters to gauge their effect. Tests of this nature are viewed as **standard series** or **methodical series**. Where standard series data of a similar hull form are available, a designer can, to some extent, interpolate within the series to obtain an estimate of resistance.

Clearly, a lot of work will be involved in a series where numerous values of different variables are used. Admiral DW Taylor of the United States, in his standard series, varied prismatic coefficient, displacement to length ratio and beam to draught ratio and tested a total of 80 models of twin-screw ships. The main parameters examined were:
- Speed coefficient (range 0.30-2.0) $= \dfrac{V}{\sqrt{L}}$

where V is speed, in knots, and L is length, in feet.

- Prismatic coefficient (range 0.48 -0.86) $= \dfrac{\nabla}{A_m L}$

where ∇ is underwater volume of displacement, in ft³, and A is underwater cross-sectional area at midships, in ft².

- Displacement:length ratio (range 20-250) $= \dfrac{\Delta}{(L/100)^3}$
  where Δ is displacement, in tons.

- Beam:draught ratio (2.5, 3.0 and 3.75).

Data on frictional and residuary resistance is presented as a series of continuous curves. The Taylor standard series data was re-analysed by Gertler.

The British Ship Research Association (BSRA), now British Maritime Technology, standard series deals with single-screw ships with varying hull proportions. The main parameters examined were:
- Block coefficient (range 0.65-0.80) = ∇/LBT
- Longitudinal centre of buoyancy (1% aft-3% forward)
- Breadth:draught ratio (range 2.0-4.0) =B/T
- Length:displacement ratio (range 4.0-6.5) = $L/\nabla^{1/3}$

Data is presented as curves of total hull resistance, using a Froude constant notation.

The United States Maritime Administration sponsored the MarAd series which examines full-form ships with high block coefficient values, small length-to-beam ratio and large beam-to-draught ratio. Other series have been examined for high-speed displacement ships and specialist vessels such as trawlers.

A number of ship form features will affect resistance and some have been examined in the various standard series examined. Form features will largely affect wave-making resistance, since frictional resistance is largely related to the hull's wetted surface area. However, neither type of resistance can be considered in isolation and any particular form change considered may affect others.

8.6 Full scale resistance testing

William Froude carried out the first full-scale resistance tests using HMS *Greyhound*, a vessel 52.6m (172ft 6in) long by 10.1m (33ft 2in) beam, which was towed by HMS *Active* using a 58m (190ft) tow rope. Resistance was measured by a dynamometer mounted on *Greyhound* and speeds were varied from 3 to 12kts.

The results obtained were compared with those from a model of *Greyhound* and showed general agreement in form although the full-scale results gave higher values of resistance. This was attributed to hull roughness. Froude was able to conclude that his law of comparison had been verified.

The then British Ship Research Association conducted full-scale tests on a former paddle steamer, *Lucy Ashton*, in the late 1940s.[1] *Lucy Ashton* was 58.1m (190ft 6in) long by 6.4m (21ft) breadth and was tested at a draught of 1.6m (5ft 4in). Gas turbine engines were mounted above the deck so as not to affect water flow around the hull and speeds were varied from 5 to 15kts. Various hull surface conditions were tested including the use of different paints. The full-scale results were compared with tests on models of 2.7, 3.7, 4.9, 6.1, 7.3 and 9.1m (9, 12, 16, 24 and 30ft) in length.

Ship resistance estimates were made using a number of skin friction formulae and the Schoenherr formula gave the best correlation. A plot of frictional resistance coefficient, C_F, against Reynolds number, R_n, was made for the ship and model results, see Fig 8.4. The Schoenherr line was then superimposed and the results were seen to lie on a line parallel to it. Results from the larger models were corrected for tank boundary conditions and shallow water effects, which placed them more precisely on

Merchant Ship Naval Architecture

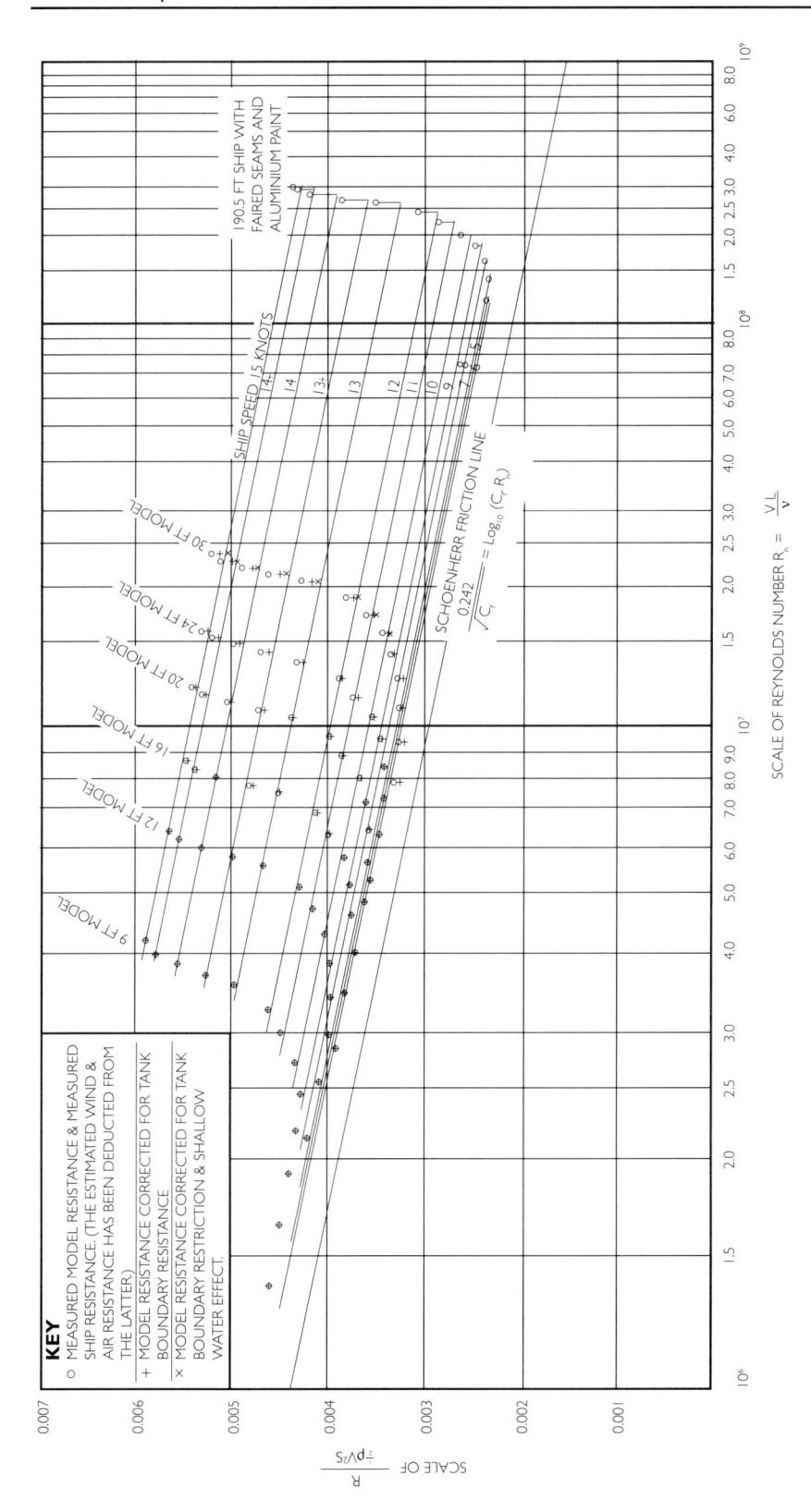

Fig 8.4: Lucy Ashton resistance results

the line. The full-scale ship results at the ship's Reynolds number gave slightly higher resistance values. This discrepancy was reduced with faired seams and the use of smooth aluminium paint.

Other full-scale tests have been conducted but the expense involved prohibits their widespread use. The tests conducted to date have shown that there is good correlation between model testing and the propulsion performance of the finished ship, making it the method of choice.

8.7 References

1. Conn, JFC, Lackenby, H, and Walker, WB, (1953). *BSRA resistance experiments on the Lucy Ashton, Part II; the ship-model correlation for the naked hull condition,* Trans RINA.

9 Propellers and propulsion

The propeller is a machine that converts the propulsion engine's power into a thrust to propel the ship at a suitable speed. The theory relating to this power conversion is discussed, and also the use of models to determine full-size propeller performance. The design procedure is detailed, and then testing and propeller problems are reviewed. The conduct of speed trials, which are used to verify both ship and propeller design, is then outlined. Finally, a number of alternative propulsion systems are reviewed.

9.1 Propellers

The screw propeller consists of a boss to which a number of blades are attached either permanently, in a fixed pitch propeller, or so they can move, in a controllable pitch propeller. The surface of a propeller blade, when viewed from behind the ship, is called the **face** and the other surface is the **back**. The face of a propeller blade forms part of a helicoidal surface and can be compared to a helical screw when turning, with each blade forming its own 'thread'. A right-handed propeller rotates in a clockwise direction when viewed from aft. Various terms are used to describe a propeller, see Fig 9.1, which will now be considered in detail.

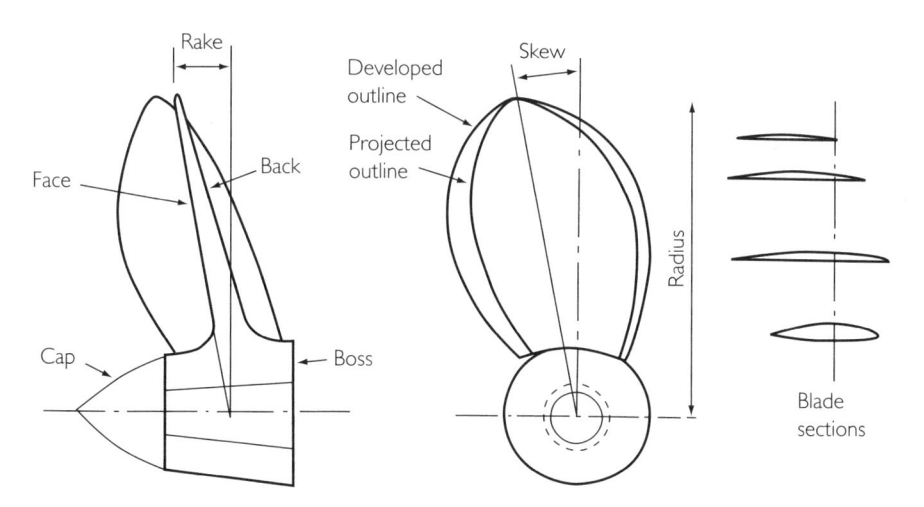

Fig 9.1: Drawing of propeller showing terminology

9.1.1 Diameter

The propeller diameter, D, is the diameter of the circle swept out, or created, by the rotating blades. It is usually made as large as possible. The boss diameter is the maximum diameter of the tapering boss to which the blades are attached at their roots. The boss diameter should be as small as practicably possible.

9.1.2 Pitch

Pitch, P, is defined as the forward distance moved by a propeller moving in an unyielding fluid in one revolution of the shaft. The face of a propeller blade forms part of a helix or helicoidal surface. A helicoidal surface is formed by a straight line when one end moves with uniform speed along an axis, while the line itself rotates with a uniform angular velocity about the axis. The distance moved along the axis in one revolution is the pitch. The pitch angle, θ, determines the pitch of the blade, see Fig 9.2, and is given by the relationship: $\tan\theta = P/\pi D$

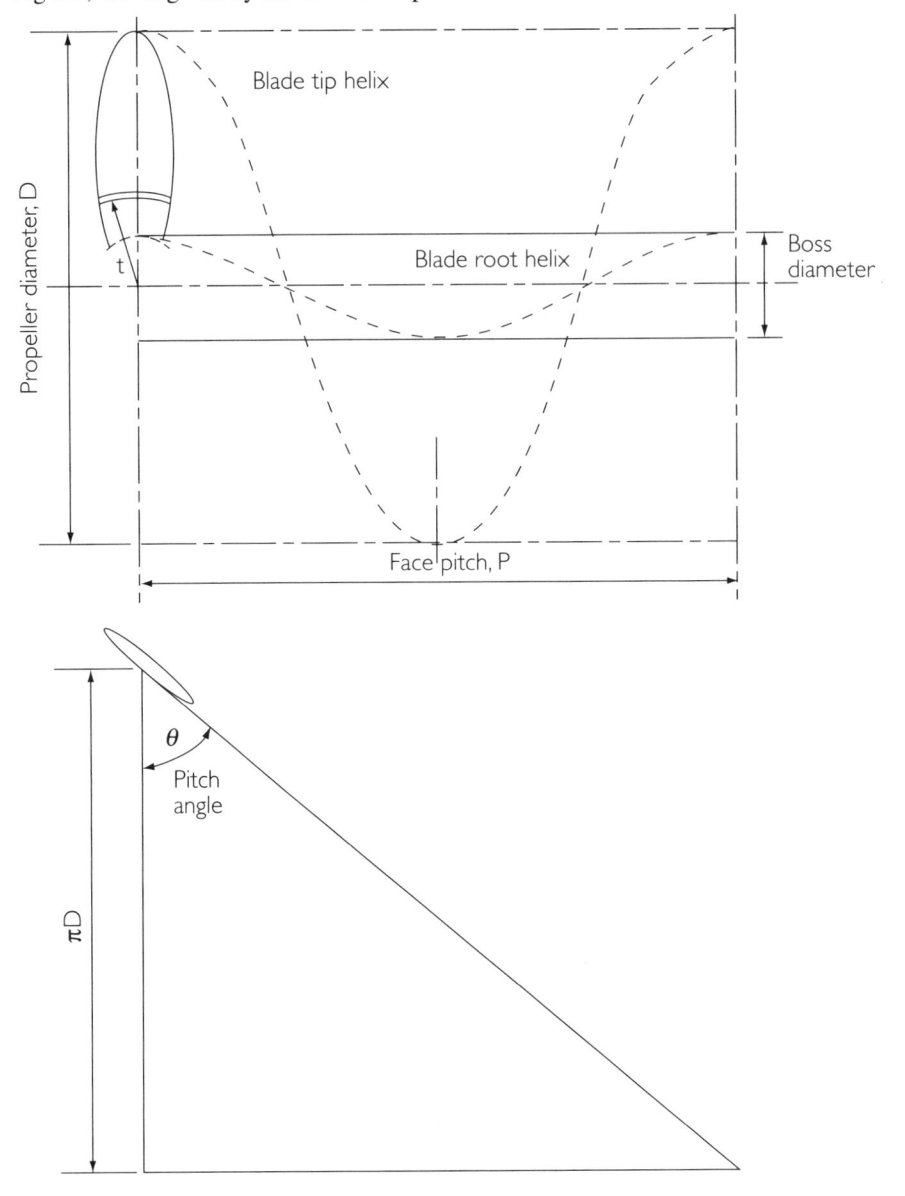

Fig 9.2: Pitch and pitch angle

In modern propellers the pitch varies along the radius, but is a fixed value at any point, such that a solid propeller can be described as a fixed pitch propeller. A **controllable pitch propeller** is one where the pitch of all the blades can be changed at the same time and by the same amount, using a mechanism within the boss, which is controlled from the ship's wheelhouse.

9.1.3 Pitch ratio
The pitch ratio, p, is the pitch divided by the diameter and is important when considering propeller performance. If the pitch varies over the blade length, then pitch ratio has to be defined in relation to a particular position on the blade, often at 0.7 of the radius.

9.1.4 Blades
The blade outline, area, section and number are all important aspects of a propeller. Since a propeller blade is not flat, its outline may be described as projected or developed. A **projected outline** is the shape seen when viewed along the axis of the propeller on a plane at right angles to it. A **developed outline** is created by marking the circumferential distance at a series of radii.

The blade area may relate to the projected or developed outline. The developed area of all blades is usually expressed as a ratio of the area represented by a circle of propeller diameter. Thus:

$$\text{blade area ratio (developed)} = \frac{\text{developed blade area}}{\text{disc area}}$$

A blade section relates to a radial section through the blade which is then laid out flat. Older propeller blades had a flat face and circular back. Modern propeller blades use aerofoil sections with a thick leading edge (meeting the fluid) and a thin trailing edge. The face of the blade may also be curved, resulting in the leading and trailing edges being raised up. The blade thickness ratio is given by:

$$\text{blade thickness ratio} = \frac{\text{maximum thickness}}{\text{blade breadth chord}}$$

The **chord** is the distance between the leading and trailing edges of the blade. The number of blades on a propeller is, to some extent, determined by the thrust loading and blade area. Single-screw merchant ships often have four-bladed propellers, while twin-screw vessels may have three-bladed propellers. As many as 11 blades have been fitted to a single propeller.

9.1.5 Rake and skew
Propeller blades are normally raked or inclined aft from the vertical, in order to improve clearance between the blade tip and the hull. Skew is the offset of the blade from the vertical in the plane of rotation. It is always a distance in the opposite direction to rotation.

9.1.6 Cap
A cone is fitted onto the after end of a propeller boss to enable a smooth flow of water from the propeller.

9.2 Propeller action theories
A propeller has from three to seven or more blades of, usually, aerofoil section mounted symmetrically on a boss, which is secured to a propeller shaft. The shaft passes through a stern tube bearing and sealing arrangement in the ship's hull and is ultimately connected to a prime mover, often a diesel engine. The forward motion of the ship and the rotation of the propeller by the engine, combine to produce relative motion between the propeller and the water. As the propeller blades rotate they create a helix, which leads to the visualisation of the propeller screwing its way through the water.

With the ship's hull in front of it, the propeller works in a disturbed water stream and effectively pushes the ship forward. A propeller is about 60 to 65% efficient in its conversion of the engine power into forward thrust in an open water situation. Astern thrust is achieved either by reversing the direction of rotation of the propeller or, if it is a controllable pitch propeller, by changing the angular setting of the blades to astern pitch.

Two theories of propeller action will now be considered, the momentum theory and the blade element theory. While both have been, to a large extent, overtaken by more recent research, they nevertheless give useful insights into thrust development and efficiency of propellers.

9.2.1 Momentum theory

The steady flow of a fluid through a screw propeller can be modelled in a simplified way by replacing the propeller with a thin actuator disc. The disc has the same diameter as the propeller, but is considered to have infinitely small thickness. All of the engine power is absorbed by the disc and dissipated as a jump in pressure, or increase in total head, of the fluid across the two faces. The fluid is considered to be ideal and thus there are no energy losses due to frictional drag.

The overall behaviour of this artificial propulsive device will be examined, rather than the detailed mechanism by which it generates thrust. The magnitude of the thrust developed will be explained in terms of changes in axial momentum within the fluid. While initially suggested by Rankine, this theory was further developed by RE Froude.

The points 1 and 3, in Fig 9.3, are considered to be an infinite distance upstream and downstream of the propeller respectively. The propeller has been replaced by an actuator disc of area, A, which is working within a stream tube at point 2. The static pressure in the slipstream at 1 and 2 will be the local static pressure p_0. The increase in pressure occurring behind the disc will be δp as also shown in Fig 9.3.

The water upstream of the propeller will be initially at rest, and will achieve a velocity at the disc of V_2 and then a velocity V_3 at some infinite distance behind it. The propeller will advance at a velocity V_1, relative to the still water. If velocities V_2 and V_3 are considered relative to the disc, then V_1 must be the velocity of advance of the disc, V_A.

The power absorbed, P_a, by the propeller is equal to the increase in kinetic energy of the slipstream per unit time thus:

$$P_a = \frac{\dot{m}}{2}\left[V_3^2 - V_1^2\right]$$

The thrust generated is equal to the increase in axial momentum of the slipstream, thus:

$$T = \dot{m}\left[V_3 - V_1\right]$$

where, in each case, $\dot{m}$ is the mass flow per unit time through the disc.

Combining the above two equations gives:

$$P_a = \tfrac{1}{2}T[V_3 + V_1]$$

This power absorbed by the propeller can also be expressed as the work done by the thrust force of the propeller:

$$P_a = TV_2$$

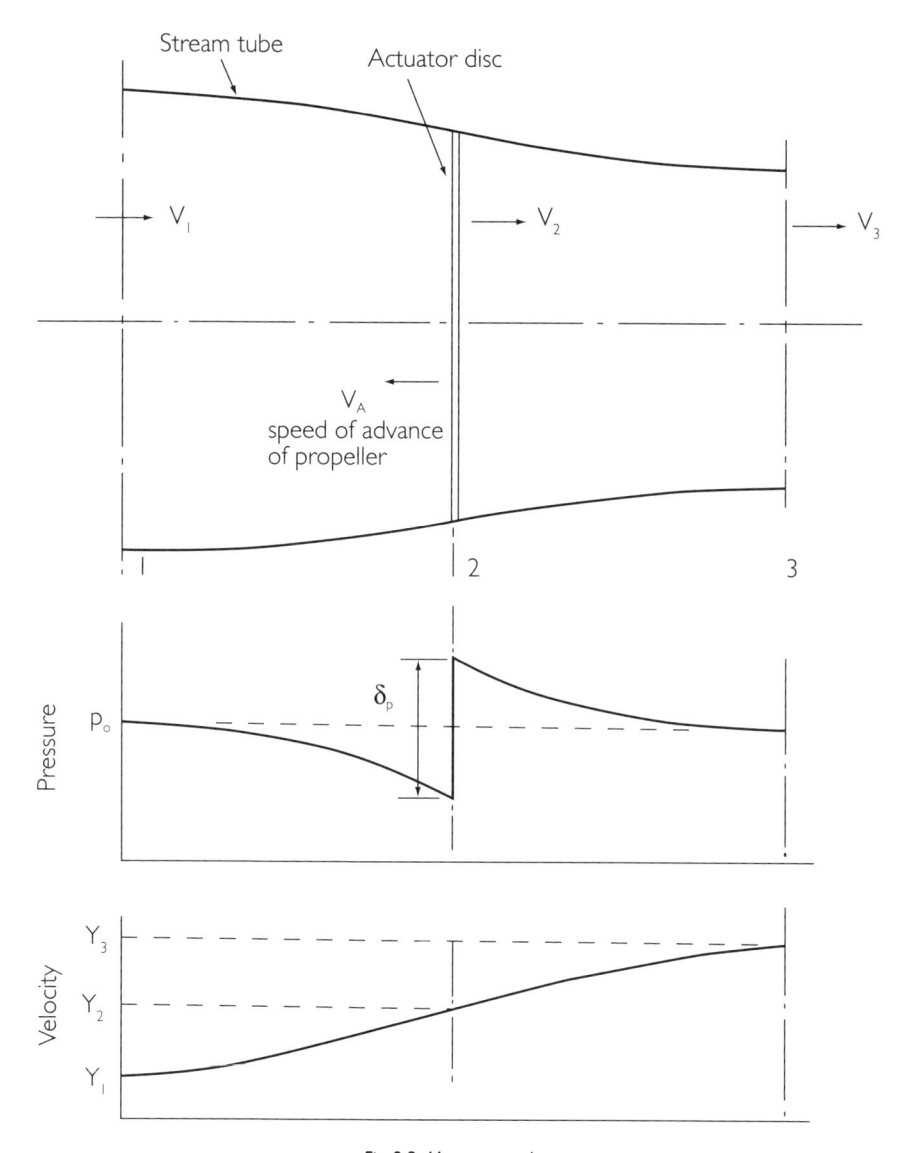

Fig 9.3: Momentum theory

Equating the two expressions for P_a gives:

$V_2 = \frac{1}{2}[V_3 + V_1]$

The velocity at the propeller disc is, therefore, the mean of the velocities at infinite distances upstream and downstream from the propeller. The velocities of the water can also be expressed in terms of the velocity at an infinite distance upstream, V_1, thus:

$V_2 = V_1 + a_2 V_1$
$V_3 = V_1 + a_3 V_1$

where a_2 and a_3 are the **axial inflow factors** at the propeller disc and at an infinite distance downstream, respectively. Substituting these two expressions into the earlier expression for V_2 gives:

$$a_3 = 2a_2$$

Thus it can be seen that half the flow acceleration takes place before the propeller disc and half after the disc. The propeller clearly has an effect on the water flow both ahead and astern. This will include the water flow around the hull and will affect hull resistance.

The useful work done by the propeller is the product of the thrust and its forward velocity V_A. Thus:

$$\text{useful work} = \text{thrust} \times V_A = \dot{m}[V_3 - V_A] \times V_A = \rho A V_2 [V_3 - V_A] \times V_A$$
$$= \rho A V_A (1 + a_2) a_3 V_A \times V_A = \rho A V_A^3 (1 + a_2) a_3$$

where ρ is the water density.

The total work done is the sum of the useful work done plus the work done in accelerating the water. The work done in accelerating the water will be the product of thrust and V_2.

The efficiency of this propeller, or disc, is the useful work done divided by the total work. Thus:

$$\text{ideal efficiency, } \eta_i = \frac{\text{useful work}}{\text{total work}} = \frac{\rho A V_A^3 (1+a_2) a_3}{\rho A V_A^3 \left[(1+a_2) a_2 a_3 + (1+a_2) a_3\right]}$$

$$= \frac{1}{1+a_2}$$

This is known as the **ideal efficiency**, η_i, of a propeller and, for a high value of η_i, the term, a_2, must be small. Thus, for a specified speed of advance, V_A, and thrust, T, the propeller diameter must be large, since $A(1 + a_2) a_3$ is constant. With a large diameter and a specified thrust the velocity of the water can be less, which results in more energy being used to propel the ship.

In this simplified approach, it has been assumed that the water has been flowing with only axial velocity, when clearly there will be some angular velocity as the water passes through the propeller blades. If this is taken into account, then the overall efficiency would be:

$$\text{overall ideal efficiency} = \frac{1-a'}{1+a_2}$$

where a' is known as the **rotational inflow factor** and will further reduce the propeller efficiency.

9.2.2 Blade element theory

The blade element theory considers the mechanism that creates forces on the blades. Account is also taken of the increase of axial and rotational velocity which occurs at the blade as outlined in the momentum theory. The blade is considered divided up into a large number of elemental strips as shown in Fig 9.4(a), each of which can be viewed as an aerofoil with water incident upon it, see Fig 9.4(b).

The aerofoil is completely immersed in water and, by reference to the relationships established in the previous chapter on resistance, an expression for the force on the aerofoil can be stated as:

$$\text{force} = \rho V^2 L^2 \left[f\left(\frac{v}{VL}\right) \right]$$

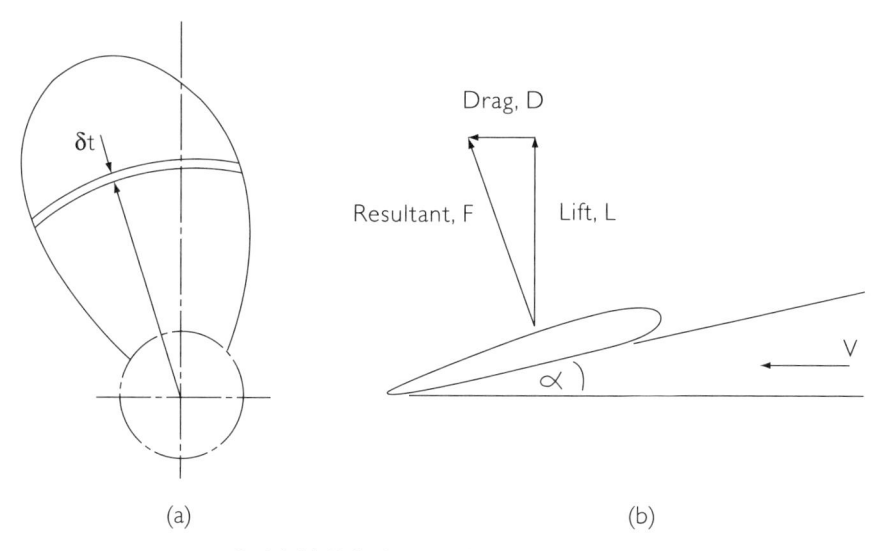

Fig 9.4: (a) Blade element (b) Forces on blade element

where ρ is density, V is speed of water flow and L is length. Now, if A is the area of one surface of the aerofoil section, then it can replace L^2 and the expression can be written in a non-dimensional form:

$$\frac{\text{force}}{\rho V^2 A} = [f(R_n)]$$

where ρ is density, V is speed of water flow, L is length and R_n is Reynolds' number.

The angle of incidence, α, of the water flow to the aerofoil section must also be taken into account, requiring a modification to the expression as follows:

$$\frac{\text{force}}{\rho V^2 A} = [f(R_n)\alpha]$$

The force on the aerofoil can be resolved into components of lift, L, which is at right angles to the direction of water flow, and drag, D, which is tangential. Coefficients C_L and C_D can be used to represent the two component forces of lift and drag respectively, where :

$$C_L = \frac{L}{\frac{1}{2}\rho V^2 A} \quad \text{and} \quad C_D = \frac{D}{\frac{1}{2}\rho V^2 A}$$

A more careful consideration of the velocity of the water incident upon the aerofoil section of the blade is now required, see Fig 9.5. The rotational velocity of the blade is 2πNr, where r is the radius of the blade element considered. In an ideal fluid, the propeller would advance in an axial direction a distance NP where P is the pitch of the blade and N is the number of revolutions per unit time since speed V has not been mentioned in unit terms. Water is not an ideal fluid and, therefore, the blade advances a lesser distance, V_A, the speed of advance, in unit time.

So far it has been assumed that the aerofoil is moving in relation to still water. However, when considering the momentum theory, it was seen that the water received an axial increase in velocity, aV_A, and a rotational increase in velocity of a'2πNr. The effect of these two factors is to increase the axial distance to $V_A(1 + a_2)$ and reduce the rotational distance to 2πNr (1- a')

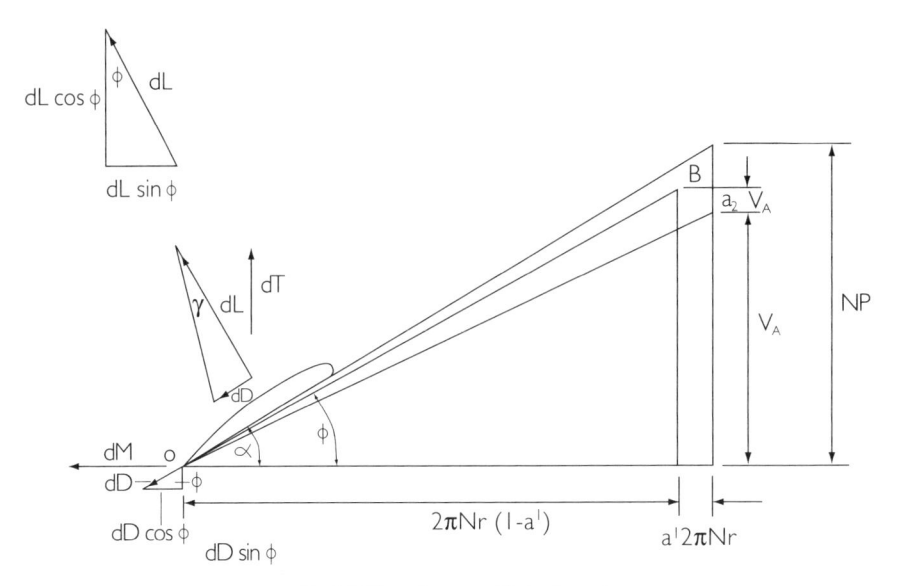

Fig 9.5: Water flow on a blade element

The water flow relative to the blade is thus along the line OB at an angle of incidence, ϕ. The element of thrust on the aerofoil will be the sum of the components of lift and drag in an axial direction, thus:

$dT = dL\cos\phi - dD\sin\phi$

The element of transverse force, dM, will be:

$dM = dL\sin\phi + dD\cos\phi$

The element of torque, dQ, will be rdM, where r is the radius of the blade element considered.

The efficiency of the blade section, $\eta = \dfrac{V_A dT}{2\pi N dQ}$

$$= \frac{V_A (dL\cos\phi - dD\sin\phi)}{2\pi Nr (dL\sin\phi + dD\cos\phi)}$$

$$= \frac{V_A dL (\cos\phi - dD/dL \sin\phi)}{2\pi Nr\, dL (\cos\phi + dD/dL \sin\phi)}$$

Considering the triangle of forces made up of the elements of lift and drag and their resultant, then $dD/dL = \tan\gamma$, and substituting gives:

$$\eta = \frac{V_A (\cos\phi - \tan\gamma \sin\phi)}{2\pi Nr (\cos\phi + \tan\gamma \sin\phi)}$$

$$= \frac{V_A}{2\pi Nr} \times \frac{1}{\tan(\phi + \gamma)}$$

The expression $V_A/2\pi Nr$ can be rewritten as:

$$\frac{V_A}{2\pi Nr} = \frac{V_A(1+a_2)}{2\pi Nr(1-a')} \times \frac{(1-a')}{(1+a_2)}$$

By reference to Fig 9.5 this expression can be written as:

$$\tan\phi \times \frac{(1-a')}{(1+a_2)}$$

Combining this expression with the earlier equation for the efficiency of the blade element gives:

$$\text{efficiency of blade element, } \eta = \frac{\tan\phi}{\tan(\phi+\gamma)} \times \frac{(1-a')}{(1+a_2)}$$

This expression shows that, in addition to the factor, $(1-a')/(1+a_2)$, introduced by momentum theory, efficiency is also related to the characteristics of the blade section defined by the angles ϕ and γ. It can be seen that if γ were zero, ie there was zero drag, then the efficiency would be that determined from the momentum theory. Drag clearly leads to a further loss of efficiency.

The actual calculation for a complete propeller would be complicated by the need to determine a_2 and a' for each aerofoil section considered. In addition, this simple analysis has not considered the interference effects between a finite number of blades, the effect of water flow between the blades and also water flow over the blade tips.

The momentum and blade element theories were developed towards the end of the nineteenth century. Propeller theory has advanced significantly over the years, with LC Burrill developing an analysis procedure using a method which combined momentum and blade element theories in the 1940s. Subsequent research work focused on the **lifting line** and later the **lifting surface**, concepts drawn from aerodynamic theory.

The **vortex lattice** method of analysis was developed from the lifting surface method in the late 1950's to early 60's. Various difficulties with the lifting surface analysis were resolved with the development of the **boundary element** approach in the early 1980s. The current approaches to propeller design and analysis make significant use of powerful computer programs. While these modern theories and methods of analysis show good agreement between theoretical and experimental results for pressure distributions over the blades and open water characteristics, more work remains to be done.

9.3 Propeller and model correlation

Models are used to predict the performance of propellers in a similar way to the use of ship models to predict full-size ship resistance. The criteria which make these predictions possible are based upon the laws of fluid dynamics and some assumptions, which will now be outlined with the aid of dimensional analysis.

The thrust, T, developed by a propeller can be considered to be dependent upon a linear dimension, its diameter D, the speed of advance, V_A, the revolutions per unit time, N, the density of the water, ρ, and its viscosity, μ, the static pressure of the fluid, p_0, and acceleration due to gravity, g.

Thrust can be written as a functional relationship of these various physical variables. Dimensional formulae can be used and, by equating indices, the following relationship can be deduced:

$$\text{thrust, } T = \rho V_A^2 D^2 \left[f_1\left(\frac{ND}{V_A}\right), f_2\left(\frac{\nu}{V_A D}\right), f_3\left(\frac{gD}{V_A^2}\right), f_4\left(\frac{p}{\rho V_A^2}\right) \right]$$

where ν is the kinematic coefficient of viscosity $= (\mu/\rho)$.

The term $f_2(\nu/V_A D)$ is a function of Reynolds number and, since viscous forces will not be large in relation to the total force, this term can be neglected. The term $f_3(gD/V_A^2)$ is in the form of a Froude number that relates to gravitational effects, which can also be ignored. The term $f_4(p_0/\rho V_A^2)$ deals with the static pressure and dynamic pressure of the fluid and relates to cavitation. If the propeller is considered to be operating under non-cavitating conditions, this term can be ignored for now, and will be considered separately later, under the topic of cavitation.

The term $f_1(ND/V_A)$ is a function of slip, or the advance coefficient, J. Referring back to Fig 9.5, the difference between the theoretical pitch distance, NP, and the speed of advance, V_A, expressed as a ratio of NP, can be described as **slip** or, in relation to the advance coefficient, J, since:

$$\text{slip } = \frac{NP - V_A}{NP} = 1 - \frac{V_A}{NP} = 1 - \frac{V_A}{NDp} = 1 - \frac{J}{p} \quad \text{where } p \text{ is the pitch ratio.}$$

Thus for a propeller operating well below the water's surface, in non-cavitating conditions, thrust can be expressed as:

$$\text{thrust, } T = \rho V_A^2 D^2 \times f_1\left(\frac{ND}{V_A}\right)$$

It is now possible to consider the factors necessary for correlation between two propellers, or a propeller and a model. If subscripts of s, for full size, and m, for model, are used, and the propellers are operating at the same advance coefficient and Froude number, then:

$$\frac{T_s}{T_m} = \frac{\rho_s}{\rho_m} \times \frac{V_{As}^2}{V_{Am}^2} \times \frac{D_s^2}{D_m^2}$$

Since the advance coefficients are the same then:

$$\frac{N_s}{N_m} = \frac{V_{As}}{V_{Am}} \times \frac{D_m}{D_s}$$

and for similar Froude numbers then:

$$\frac{gD_s}{V_{As}^2} = \frac{gD_m}{V_{Am}^2}$$

Combining these two expressions and using Δ to represent the ratio of linear dimensions D_s/D_m gives:

$$T_s = \frac{\rho_s}{\rho_m} \times \Delta^3 \times T_m \qquad N_m = N_s \sqrt{\Delta}$$

It should be noted that the model propeller will rotate faster than the full-scale ship's propeller. Additional relationships can be developed and these are listed in Table 9.1

Factor	Full size propeller	Model propeller
Diameter, D	$D = \Delta d$	d
Speed of advance, V_A	$V_A = \Delta^{1/2} v$	v
Pitch, P	$P = \Delta p$	p
Revolutions per minute, N	N	$n = \Delta^{1/2} N$
Advance coefficient, J	J	J
Thrust, T	$T = \Delta^3 t$	t
Torque, Q	$Q = \Delta^4 q$	q
Thrust power, T_P	$T_P = \Delta^{3.5} tv$	tv
Delivered power, D_P	$D_P = \Delta^{3.5} 2\pi n q$	$2\pi n q$
Efficiency, η	$\eta = \dfrac{T_P}{D_P} = \dfrac{K_T}{K_Q} \times \dfrac{J}{2\pi}$	$\dfrac{tv}{2\pi n q}$
Disc area, A	$A = \Delta^2 a$	a

Table 9.1: Propeller and model relationships

Example

A model propeller of 0.41m diameter with a pitch ratio of 1.25, produces thrust of 266N, when advancing at a speed of 6.5kts at 800rev/min with a torque of 16.4Nm. Determine, for a geometrically-similar ship's propeller of 4.3m diameter, the pitch, rev/min, thrust, torque, thrust power, delivered power and efficiency.

Since pitch ratio = P/D = 1.25
Model propeller pitch, P = 1.25 × 0.41 = 0.512m
Dimensional ratio, Δ = 4.3/0.41 = 10.5

For full size ship propeller:
Pitch, P = Δp = 10.5 × 0.512 = 5.37m

Rotational speed $= \dfrac{n}{\Delta^{\frac{1}{2}}} = \dfrac{800}{\sqrt{10.5}} = 247 \text{rev/min}$

Thrust, $T = \Delta^3 t = 10.5^3 \times 266 = 308\,000\text{N} = 308\text{kN}$

Torque, $Q = \Delta^4 q = 10.5^4 \times 16.4 = 199\,343\text{Nm}$

Speed of advance $V_A = \Delta^{1/2} v = \sqrt{10.5} \times 6.5 = 21$ knots

Thrust Power, $T_P = T \times V \times 0.5144 = 308 \times 21 \times 0.5144 = 3327\text{kW}$

Delivered Power, $D_P = \dfrac{2\pi QN}{1000} = 2\pi \times 199\,343 \times \dfrac{247}{60} \times \dfrac{1}{1000} = 5156\text{kW}$

Efficiency, $\eta = \dfrac{T_P}{D_P} = \dfrac{3327}{5156} = 64.5\%$

9.4 Propeller coefficients

An expression for thrust was determined earlier, where:

thrust, $T = \rho V_A^2 D^2 \times f_1 \left(\dfrac{ND}{V_A} \right)$

and the advance coefficient, $J = ND/V_A$. Substituting for V_A gives;

$$K_T = \frac{T}{\rho N^2 D^4}$$

where K_T is a non-dimensional thrust coefficient, which will have the same value for geometrically-similar propellers working at the same advance coefficient.

Similarly, K_Q is a non-dimensional torque coefficient, where :

$$K_Q = \frac{Q}{\rho N^2 D^5}$$

In order to be non-dimensional, the various terms must have the following units: T = Newtons; Q = Newton metres; N = rev/sec; D = metres; V_A = metres/sec; $\rho = 1025 \text{kg/m}^3$.

Propeller efficiency can be expressed as:

$$\eta = \frac{\text{useful output}}{\text{input}} = \frac{TV_A}{Q \times 2\pi N} = \frac{K_T}{K_Q} \times \frac{J}{2\pi}$$

The thrust and torque coefficients, together with efficiency, can be plotted against advance coefficient to give a chart as shown in Fig 9.6.

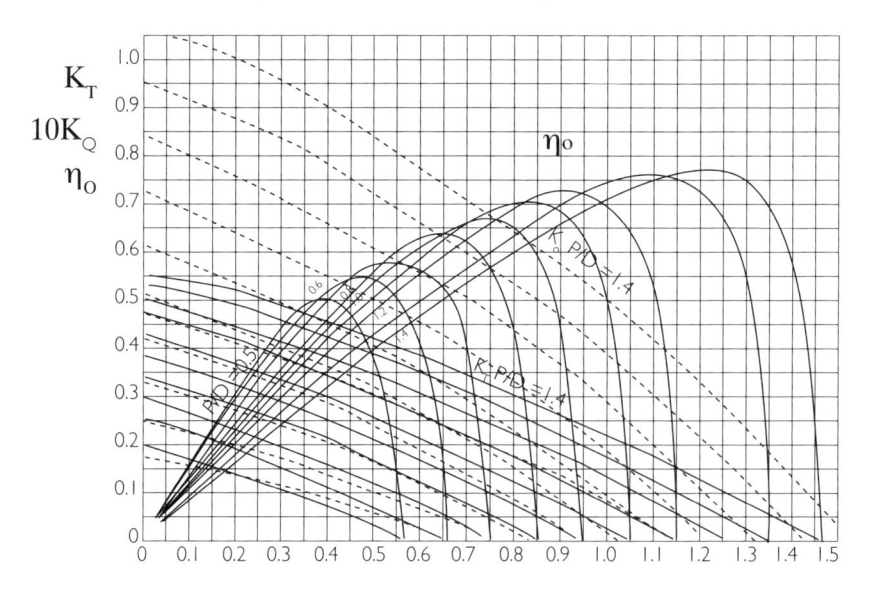

Fig 9.6: Curves of thrust, torque and efficiency

Example
The following data is provided for a propeller:
Diameter = 4.88 m; K_Q = 0.015; K_T = 0.097; J = 0.633; N = 120rev/min.
Determine the values of delivered power, P_D, thrust power, P_T, speed of advance, V_A and efficiency, η.

From $K_Q = \dfrac{Q}{\rho N^2 D^5}$

torque, $Q = K_Q \rho P N^2 D^5$

$$= 0.015 \times 1025 \times \left(\frac{120}{60}\right)^2 \times 4.88^5$$

$$= 170\ 200 \text{Nm}$$

$$P_D = \frac{2\pi QN}{1000}$$

$$= 2\pi \times 170\ 200 \times \frac{120}{60} \times \frac{1}{1000}$$

$$= 2140 \text{kW}$$

From $K_T = \dfrac{T}{\rho N^2 D^4}$,

thrust, $T = K_T \rho N^2 D^4$

$$= 0.097 \times 1025 \times \left(\frac{120}{60}\right)^2 \times 4.88^4$$

$$= 225\ 500$$

$$= 225.5 \text{kN}$$

$$J = \frac{V_A}{ND}$$

$$V_A = JND = 0.633 \times \left(\frac{120}{60}\right) \times 4.88$$

$$= 6.18 \text{m/sec}$$

$P_T = T \times V_A$
$= 225.5 \times 6.18$
$= 1390 \text{kW}$

efficiency, $\eta = \dfrac{P_T}{P_D} = \dfrac{1390}{2140} = 65\%$

9.5 Propeller design

When designing a propeller, the procedure is to select a diameter and pitch that will provide a desired power at some stated value of revolutions per minute and ship speed. Model test data for design purposes are derived from systematic tests on groups of propeller models which differ, generally, in only one particular at a time. The basic variable for each group is the pitch ratio, p, where $p = P/D$ where P is propeller pitch and D is diameter.

Early work on model propellers was carried out by RE Froude, but data is now generally presented in the form of Taylor constants, named after DW Taylor. These constants were developed from the expression for thrust:

thrust, $T = \rho V_A^2 D^2 \times f_1\left(\dfrac{ND}{V_A}\right)$

and by substituting for $D = V_A/JN$

thrust, $T = \dfrac{\rho V_A^4}{N^2} \times \Psi(J)$

where ψ is some new function of J which includes J^2.

Now Taylor used U to represent thrust power $= TV_A$, thus:

$U = TV_A = \dfrac{\rho V_A^4}{N^2} \times \Psi(J) \times V_A = \dfrac{\rho V_A^5}{N^2} \times \Psi(J)$

The Taylor constant for thrust power, B_U, was obtained by considering ρ to be for sea water and using the square root of the remaining terms:

$B_U = NU^{0.5}/V_A^{2.5}$

Other coefficients used by Taylor are:

$B_P = NP^{0.5}/V_A^{2.5}$ where P is shaft power.

$\delta = ND/V_A$ where δ is the reciprocal of advance coefficient.

$\eta = P_T/P_D$ where η is efficiency.

In the original work by Taylor, diameter, D, was in feet, speed, V_A, in knots, N in revolutions per minute and power was in horsepower. The Taylor coefficients expressed in metric units are:

$B_U = 1.158NU^{0.5}/V_A^{2.5}$

$B_P = 1.158NP^{0.5}/V_A^{2.5}$ and

$\delta = 3.2808ND/V_A$

where diameter, D, is in metres, the speed, V_A, is in knots, N is in revolutions per minute, and power, P, is in kilowatts.

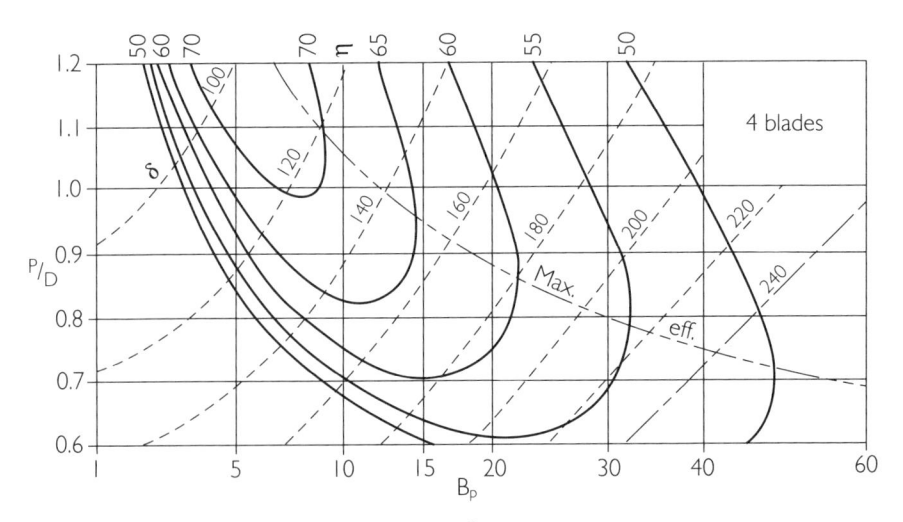

Fig 9.7: B_P-δ diagram

Taylor plotted the results for a family of propellers of the same geometrical design, but different pitch ratio on a B_P–δ diagram as shown in Fig 9.7. Curves of δ and η are plotted on a base of B_P or B_U with an ordinate of pitch ratio. If the power, speed of advance and revolutions can be estimated for a particular propeller design, the diagram will enable the best combination of diameter, pitch and revolutions to be selected. If an ordinate is erected on the diagram at the calculated value of B_P or B_U, it will be seen that there is some value of the pitch ratio above or below which efficiency falls. This is the optimum pitch ratio which should be chosen. The value of δ can thus be found and the diameter calculated and, hence, the pitch.

Example

Determine the optimum dimensions for a propeller, given the following design data: Delivered power, $P_D = 4800\text{kW}$; shaft speed, $N = 100\text{rev/min}$; speed of advance, $V_A = 11.04$ knots.

$$Bp = \frac{1.158NP^{0.5}}{V_A^{2.5}} = \frac{1.158 \times 100 \times 4800^{0.5}}{11.04^{2.5}} = 19.81$$

From Fig 9.6, when $B_P = 19.8$; $\delta = 173$; $\eta = 0.615$; $P/D = 0.885$

since $\delta = \dfrac{3.2808ND}{V_A}$ $\therefore D = \dfrac{173 \times 11.04}{3.2808 \times 100} = 5.82\text{m}$

Pitch ratio, $p = P/D = 0.885$ $\therefore P = 0.885 \times 5.82 = 5.15\text{m}$

Assume the maximum diameter is restricted to 5m

$$\delta = \frac{3.2808ND}{V_A} = \frac{3.2808 \times 100 \times 5}{11.04} = 149$$

From Fig 9.6, $p = 1.162$ $\therefore P = 1.162 \times 5 = 5.81\text{m}$

$\eta = 0.585$

$\therefore$ Loss of efficiency $= 61.5\% - 58.5\% = 3\%$

9.6 Propeller testing

The test data discussed earlier was obtained from tests carried out on model propellers in towing tanks without a ship's hull in front, which are thus known as 'open water' tests. The towing tank carriage is moved along at the required speed of advance and the propeller is mounted forward of a streamlined casing which contains the drive shaft.

The thrust, torque, revolutions per minute of the propeller and speed of the carriage are recorded and from this data K_T, K_Q, J and η can be calculated for a standard series of propellers. In a standard series only pitch ratio is varied on the different propeller models tested. This 'methodical series' data, as it is known, is used to produce B_P-δ diagrams. Notable researchers who have produced methodical series include Froude, Gawn, Taylor, Troost and Van Lammeren.

9.6.1 Propeller and ship interaction

The resistance of the ship's hull has been considered without reference to the presence of the propeller, and open water tests on propellers are undertaken without a hull in front of the ship. When the two are considered together, as they would be on a ship, then interaction effects will occur, which will now be considered.

There are three interaction effects:
1. Some of the water flowing around the hull achieves a velocity relative to it and is called the wake. As a result, the velocity of water relative to the propeller disc is not equal to the speed of advance of the propeller relative to still water.
2. Water flow across the propeller disc will vary in magnitude and direction because of the hull shape, further affecting propeller performance.
3. Propeller action will vary pressures within the water whose effect will be to increase hull resistance.

The wake velocity varies in magnitude, and also direction, but is generally considered to have a forward velocity giving a positive value:

Wake velocity = $V - V_A$ where V is the speed of the ship and V_A is the average velocity of water relative to the hull at the propeller position, ie speed of advance.

Wake speed can be expressed as a percentage of either the speed of advance, V_A, of the propeller, or the speed of the ship, V. Froude selected V_A, while Taylor selected V, thus:

$$\text{Froude wake fraction} = w_F = \frac{V - V_A}{V_A}$$

$$\text{Taylor wake fraction} = w_T = \frac{V - V_A}{V}$$

The relationship between the two is:

$$w_F = \frac{w_T}{1 - w_T} \quad \text{and} \quad w_T = \frac{w_F}{1 + w_F}$$

The variation in velocity across the propeller disc results in a lower propeller efficiency compared with open water and the term **relative rotative efficiency**, η_R, is used to take this into account.

$$\text{Relative rotative efficiency, } \eta_R = \frac{\text{propeller efficiency behind hull}}{\text{propeller efficiency in open water}}$$

The change in pressure variations on the hull brought about by propeller action were discussed earlier in relation to momentum theory. Their effect is to increase or augment hull resistance or reduce the thrust of the propeller. Two expressions are in use for the same effect:

$$\text{augment of resistance, } a = \frac{(T - R)}{R} = \frac{T}{R} - 1 \text{ or } 1 + a = \frac{T}{R}$$

$$\text{thrust deduction factor, } t = \frac{(T - R)}{T} = 1 - \frac{R}{T} \text{ or } 1 - t = \frac{R}{T}$$

Hull efficiency
The ratio of effective power to thrust power is known as the **hull efficiency**. Effective power is the power required to tow the ship of resistance, R_T, at some speed, V. Thus effective power $P_E = R_T \times V$

Thrust power, P_T, is the product of thrust developed by the propeller and the speed of advance, V_A. Hence:

hull efficiency, $\eta_H = \dfrac{P_E}{P_T} = \dfrac{R_T \times V}{T \times V_A}$

Now since $T = R_T(1 + a)$ and $V_A = \dfrac{V}{1 + w_F}$

∴ Hull efficiency, $\eta_H = \dfrac{1 + w_F}{1 + a}$ and since $(1 + a) = \dfrac{1}{(1-t)}$

$= (1 + w_F)(1 - t)$

The hull efficiency is, curiously, a little greater than unity, since the propeller benefits from some of the energy in the water which has been set in motion by the hull. The wake fraction, augment of resistance and relative rotative efficiency are known collectively as the **hull efficiency elements**.

Propeller efficiency

The propeller efficiency in open water, η_O, is the ratio of thrust power to delivered power when developing a thrust, T, in open water at a speed of advance V_A.

Propeller efficiency in open water, $\eta_O = \dfrac{P_T}{P_D}$

Overall propulsive efficiency

The shaft power, P_S, of the engine is used to propel a ship complete with appendages and propeller. The ratio of effective power, P_E, to shaft power is the overall propulsive efficiency and is called the **propulsive coefficient**.

Propulsive coefficient $= \dfrac{P_E}{P_S}$

The efficiency of the transmission between the shaft power produced by the engine and the power absorbed by the propeller, P_D, is included in this coefficient. If this is removed, then the more commonly used **quasi-propulsive coefficient** (QPC) is obtained.

Propulsive coefficient = quasi-propulsive coefficient × transmission efficiency.

Now $QPC = \dfrac{P_E}{P_D} = \dfrac{P_E}{P_E'} \times \dfrac{P_E'}{P_T} \times \dfrac{P_T}{P_D'} \times \dfrac{P_D'}{P_D}$

and

$\dfrac{P_E}{P_E'} = \dfrac{1}{\text{appendage coefficient}}$

$\dfrac{P_E'}{P_T} = $ hull efficiency, η_H,

$\dfrac{P_T}{P_D'} = $ propeller open water efficiency, η_O,

$\dfrac{P_D'}{P_D} = $ relative rotative efficiency, η_R,

Hence:

$$QPC = \left[\frac{\eta_H \times \eta_O \times \eta_R}{\text{appendage coefficient}}\right]$$

The quasi-propulsive coefficient can be determined from model tests, but in scaling-up to the full-size ship some allowance must be made for hull roughness, fouling, wind, waves, etc, and this is usually in the form of a load factor, where:

$$\text{load factor} = (1+x) = \left[\frac{\text{transmission efficiency}}{QPC \times \text{appendage coefficient}}\right]$$

The term, x, is known as the **overload fraction**.

Example

Determine the shaft power for a ship and the optimum propeller dimensions using the Bp-δ diagram in Fig 9.7 and the following design data:
Effective power naked, P_E naked = 2900kW; appendage and roughness allowance 8%; weather allowance 15%; transmission losses 3%; quasi-propulsive coefficient 0.75; ship speed = 16 knots, Taylor wake fraction = 0.31.

Effective power of ship with appendages and allowance for roughness =2900 (1 + 0.08) = 3132kW.
Effective power of ship with weather allowance = 3132 (1 + 0.15) = 3602kW
Delivered power = Effective power with allowances/QPC = 3602/0.75 = 4802kW
Shaft power = 4802(1 + 0.03) = 4947kW

Considering now the propeller design:
Speed of advance, V_A = V(1 - w_T) = 16(1 - 0.31) = 11.04 knots
Shaft rev/min will now be considered for 100, 90 and 80rev/min and the corresponding values of B_P will be calculated and used in Fig 9.7.
For N = 100rev/min

$$B_P = 1.158 \times \frac{100 \times 4800}{(11.04)^{2.5}} = 19.81$$

From Fig 9.7, when B_P = 19.81, then δ = 170, η = 0.62 and P/D = 0.89
For N = 90rev/min

$$B_P = 1.158 \times \frac{90 \times 4800}{(11.04)^{2.5}} = 17.83$$

From Fig 9.7, when B_P = 17.83, then δ = 163, η = 0.63 and P/D = 0.90
For N = 80rev/min

$$B_P = 1.158 \times \frac{80 \times 4800}{(11.04)^{2.5}} = 15.85$$

From Fig 9.7, when B_P = 15.85, then δ = 156, η = 0.64 and P/D = 0.93
It can be seen that propeller efficiency increases as the engine revolutions per minute are reduced. The actual choice of both engine shaft power and revolutions will be decided by what is available from engine manufacturers or the particular engine manufacturer of choice.

9.7 Propeller design procedure

The various stages in the propeller design procedure used in the previous example can be summarised as follows:

1. Estimate the P_E naked requirement for the ship speed required, using a model test or standard series data applicable to the ship type being considered.
2. Add to the P_E naked the allowances for appendages and weather. For appendages on a single screw ship this will be about 8% and for twin screw vessels about 12%. The weather allowance can be as high as 15%, or even more on routes such as the North Atlantic.
3. Obtain a value for the quasi-propulsive coefficient and thus determine the delivered power, P_D.
4. Obtain the wake fraction from a similar ship and determine the speed of advance.
5. Select a value for engine rev/min, probably by consulting a manufacturer's catalogue.
6. Calculate B_P and from a propeller chart obtain the value of optimum pitch ratio and δ.
7. Use these values to obtain the diameter and pitch.

9.8 Model measurement of hull efficiency

The various elements of hull efficiency can be determined for a ship model which is fitted with a motor-driven propeller. The model is moved down the test tank at the ship's corresponding speed, and the propeller rev/min is varied. The model speed, and resistance, and the propeller thrust, torque and rev/min are measured and the results plotted as shown in Fig 9.8. The model self-propulsion point can be found where the lines representing propeller thrust and model resistance with propellers cross. Separate tests of the model without a propeller will establish a base from which the **augment of resistance** for the model can be measured.

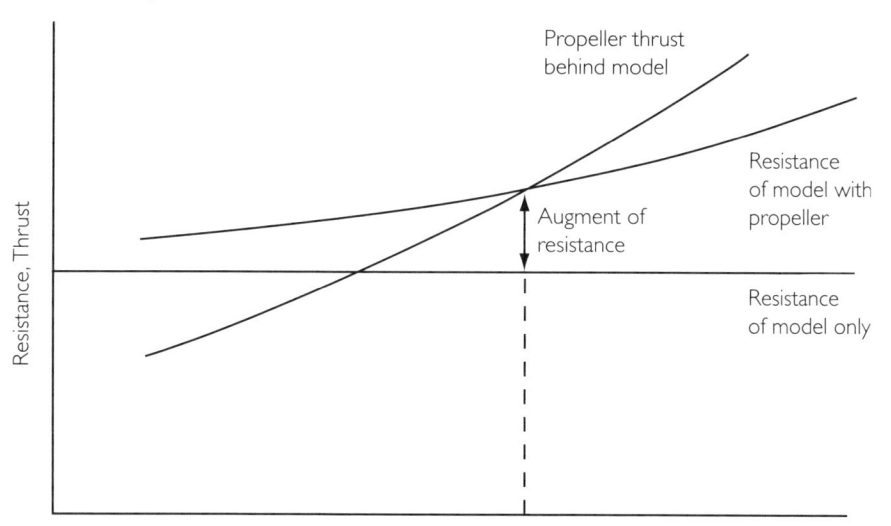

Fig 9.8: Model self propulsion point

Open-water test results for the model propeller can now be used to determine the wake speed. If a plot of thrust coefficient to advance coefficient is available, then the actual thrust to propel the model can be used to determine the value of the advance coefficient, J. Now since $J = V_A/ND$ and the values of N and D are known, the speed of advance, V_A, can be found. The difference between the model self-propulsion speed and V_A will give an average wake speed.

The relative rotative efficiency is found from the ratio of the torques measured in each of the open-water and model with propeller, tests.

The hull efficiency elements determined in this way are usually used with methodical series data when designing a propeller, since the propeller models are too small to directly scale up thrust and torque values.

9.9 Cavitation

The thrust and torque of the propeller depend upon the lift and drag characteristics of the blade sections. The lift on the section is produced partly from the suction on the back of the blade and partly from the positive pressure on the driving face. Normally the suction force is about four times greater than the pressure force.

As the propeller revolutions increase, the peak of the pressure reduction curve increases and if at any point the local pressure on the blades falls below the water vapour pressure, a cavity, filled with water vapour and air, will be formed. This phenomenon leads to a loss of thrust and torque, an increase in revolutions and, ultimately, collapse of the cavitation bubbles. This can lead to intense local pressures that can cause pitting or erosion of the blades. Pitting will weaken the propeller and also increase the surface irregularities of the blades.

Prior to the onset of cavitation, the propeller blades may give out a high-pitched note. This **singing**, as it is called, is due to vibration of the blades set off by resonant shedding of non-cavitating eddies from the trailing edge of the blades. It can be eliminated by a change in blade shape or area.

9.9.1 Ship and model cavitation correlation

The non-dimensional parameter associated with pressure, established when examining ship resistance and propeller operation, was $p_o/\rho V^2$, where p_o is the static pressure, ρ is water density and V is velocity. For dynamic similarity between the ship and model, this non-dimensional quantity must be the same for each. Using subscripts of s for ship and m for model, then:

$$\frac{p_{om}}{\rho_m V_m^2} = \frac{p_{os}}{\rho_s V_s^2}$$

Now, since the ship and model propellers must also operate at the same Froude number, then:

$$V_m = \frac{V_s}{\Delta^{0.5}}$$

where Δ is the ratio of linear dimensions L_s/L_m,

$$\therefore p_{om} = \frac{\rho_m}{\rho_s} \times \frac{p_{os}}{\Delta}$$

If the small difference in sea water and fresh water densities is ignored, then for dynamic similarity, the model testing pressure must be reduced in relation to the ratio of linear dimensions. This can readily be done for water pressure, but the atmospheric pressure requires the testing facility to be depressurised. Some testing tanks have this facility, but generally use is made of a cavitation tunnel.

9.9.2 Cavitation number

At a pressure below the vapour pressure, p_v, the air dissolved in water can be released in the form of bubbles. A **cavitation number** is used to indicate when bubbles are released and cavitation is about to occur on a propeller.

$$\text{Cavitation number, } \sigma = \frac{(p_o - p_v)}{\frac{1}{2}\rho V^2}$$

Conditions will vary across the blade and, for a standard condition, the speed is taken as the speed of advance, V_A, and the static pressure p_o is measured at the centreline of the propeller hub. At any local condition on a blade the actual values will be used.

9.9.3 Types of cavitation

Different patterns of cavitation can occur on propellers and they are usually grouped as face, sheet, bubble, tip vortex and hub vortex. The greatest pressure reductions usually occur on the back of the blade and most cavitation occurs on this side. High velocity at the blade tips lowers the cavitation number and makes its occurrence here most likely. The high angle of incidence where the blade meets the hub can also bring on cavitation.

Face cavitation occurs at the blade leading edge, usually when the angle of incidence is small. Tip vortex cavitation occurs due to the low pressure within the vortex at the blade tip. A similar situation can occur at the blade root where it joins the hub. Sheet cavitation occurs when large suction pressures build up near the leading edge of the blade and results in part of the back of the blade being covered by a sheet of bubbles.

A propeller is considered to be **fully cavitating** when the back of the blade is covered in sheet cavitation; the term **super cavitating** is also used.

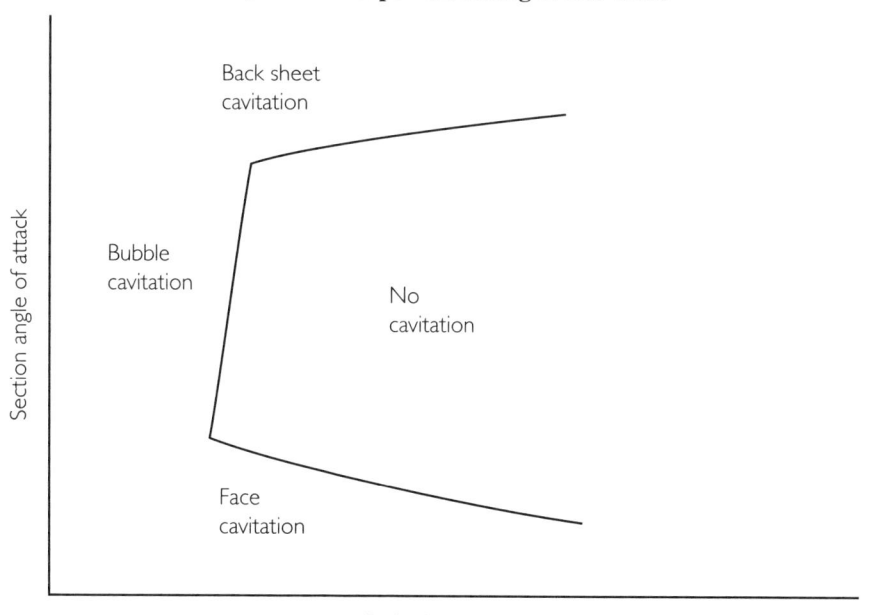

Fig 9.9: Cavitation bucket diagram

When considering propeller design in relation to cavitation, use can be made of a **cavitation bucket diagram**, see Fig 9.9. This is plotted in relation to blade section, angle of attack and blade section cavitation number. Four areas are identified in the diagram which shows where no cavitation occurs and where each of face-, bubble- and back cavitation may occur.

9.9.4 Cavitation tunnel

A fully-enclosed cavitation tunnel enables the scaling of atmospheric pressure to achieve dynamical similarity between the model and full-scale propeller. Water is circulated around a tunnel located in a vertical plane, by a large motor-driven impeller. Guide vanes and deflectors are used to provide a uniform water flow on to a model propeller in the working section of the tunnel, which is fitted with observation windows. Water pressure in the tunnel is reduced using a vacuum pump, and dissolved gas and air are removed from the water.

The propeller is usually tested in, effectively, 'open water' conditions, but in larger tunnels it is possible to fit a model hull in front of the propeller to reproduce wake variations. Stroboscopic lighting enables seemingly-stationary propeller blades to be viewed.

When testing propellers in an open water condition in the tunnel, the water flow speed represents the speed of advance and can be varied. Propeller shaft revolutions can be varied using the motor drive, and thrust and torque can be measured. The propeller can thus be tested over a range of values of advance coefficient, J, and by adjusting the static pressure in the tunnel, cavitation numbers can also be varied.

A plot of efficiency, thrust and torque coefficients against advance coefficient, for varying values of cavitation number, shows that each reduces as the J value reduces. The greater the cavitation number, the greater is the loss in efficiency.

9.10 Speed trials

A completed ship will undergo a series of tests and trials in order to satisfy the owner that all specified conditions have been met, to verify the design criteria used when predicting performance, and to provide operational information for use by the ship's officers.

Speed and power trials will now be examined, in order to consider the information in relation to propeller design and resistance estimates obtained from models. Speed trials will be run at full power to obtain maximum speed and also over a series of progressively increasing speeds, ie, **progressive speed trials**.

Measurements of ship speed, engine rev/min and power are taken during the trials. Ship speed through the water can be affected by currents, waves and weather conditions and, since it is an important contractual item, special care is taken in its measurement.

9.10.1 Ship speed measurement

The accurate measurement of ship speed on trials was traditionally made by running the ship over a measured mile course, but increasing use is now made of global positioning systems, which use satellite fixes to determine the ship's speed in relation to the land.

A measured mile is usually set up on a strip of coastline with two pairs of posts set up at a precise distance apart, ideally one nautical mile. The ship steers a course at right angles to the pairs of posts and then travels on for several miles before turning to repeat the run in the opposite direction. Usually several runs are made over the same course in each direction. In order to ensure accurate results a number of additional precautions are needed. The weather should be fine, with little or no wind, seas should be calm, less than Beaufort Scale 2-3, with low swell. The water should be deep, steady speeds should have been attained prior to reaching the measured mile and the ship should have a clean, recently-painted hull. Water density and the ship's draughts must be measured to accurately determine the displacement.

The mean speed over a series of runs is calculated by the mean of means or averaging formula:

$$\text{ship speed} = 1/8(V_1 + 3V_2 + 3V_3 + V_4) \text{ for four runs}$$
$$= 1/32(V_1 + 5V_2 + 10V_3 + 10V_4 + 5V_5 + V_6) \text{ for six runs.}$$

Similar formulae can be used to find the mean revolutions per minute of the engine.

9.10.2 Analysis of trial data

If a plot of ship speed through the water against trial power is made, the trial power predicted from model tests can be superimposed upon it. This will enable a ship-model correlation factor — ie the difference in power at any particular speed — to be determined, which will be useful in the design of future similar ships.

Where a torsionmeter has been used to measure shaft torque during a speed trial, then values of the coefficient K_Q can be calculated. If the ship speed is used to determine corresponding values of the advance coefficient, J, then a plot can be made. Open water tests of the ship's propeller will enable the advance coefficient for any particular torque coefficient to be found, which will provide a value of the speed of advance, V_A. The mean wake speed can be found as the difference between the ship's speed and the speed of advance at the value of torque coefficient chosen.

9.11 Alternative propulsion systems

Numerous alternatives to, and variations of, the fixed pitch propeller exist for ship propulsion and are in commercial use today. Propeller alternatives may be controllable pitch, fitted in a duct, or have another unit fitted behind it. Propulsion system alternatives include the cycloidal propeller, thrusters, waterjets, and sails.

9.11.1 Controllable pitch propellers

A controllable pitch propeller is made up of a boss with separate blades mounted into it. An internal mechanism enables the blades to be moved simultaneously through an arc to change the pitch angle and, therefore, the pitch. A typical arrangement is shown in Fig 9.10.

The controllable pitch propeller hub must be larger than that of a fixed pitch propeller in order to accommodate the linkages and elements of the pitch-changing mechanism. The control mechanism, which is usually hydraulic, passes through the tailshaft and operation is from the wheelhouse.

Varying the pitch will vary the thrust provided and, since a zero pitch position exists, the propulsion engine shaft may turn continuously without the ship moving through the water. The blades may be rotated to reverse the pitch and provide astern thrust. This avoids the need for a reversing gearbox or, in the case of direct drive engines, they do not need to be reversed.

9.11.2 Cycloidal propellers

Cycloidal propellers are vertical rotating devices which, in addition to propulsion, also enable the vessel to be steered. These units have blades that are vertically positioned around a disc and can be rotated by cams in order to change the angle of every blade at a particular point in each revolution. This results in a thrust whose magnitude and direction is determined by the cams. It is, therefore, in some respects similar to a controllable pitch propeller in that the disc is driven and the blades can be positioned independently of the main drive.

a) ahead pitch

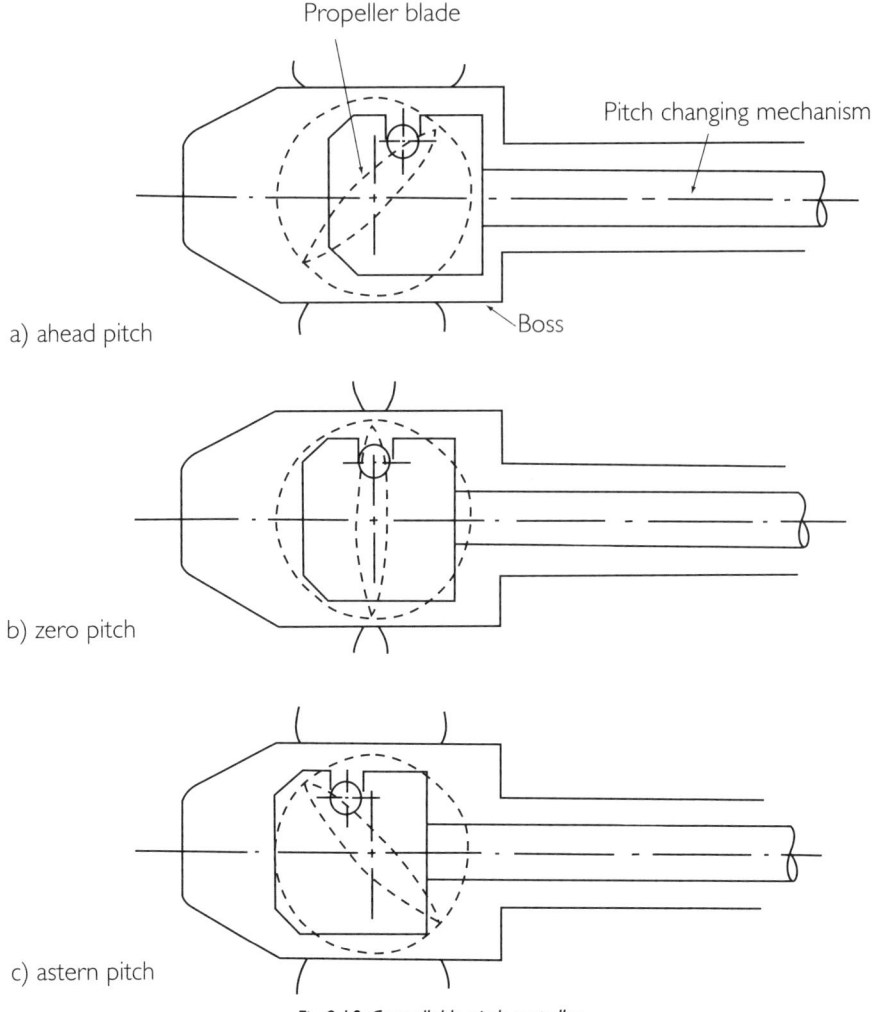

b) zero pitch

c) astern pitch

Fig 9.10: Controllable pitch propeller

This unit can effectively thrust in any direction and will respond rapidly to the pitch control mechanism. The complete assembly is unfortunately complex, can be noisy in operation and necessitates considerable maintenance. It is often used for main propulsion in ferries and other vessels requiring considerable manoeuvrability. It may also be used as a thruster or propulsion device for drill ships or floating cranes, which require accurate positioning.

9.11.3 Ducted propeller

The use of a duct or nozzle around the propeller can result in an improvement of the propeller performance. Furthermore, the aerofoil shape of the duct can produce a forward thrust that will offset any drag it causes. The duct also protects the propeller from damage and reduces noise. The propeller within the duct may be of the fixed or controllable pitch type. The ducted propeller is usually fitted on ships with heavily loaded propellers, eg tugs, but has been used on larger ocean-going vessels. One particular patented design of duct is known as the Kort Nozzle.

9.11.4 Contra-rotating propellers

The contra-rotating propeller arrangement uses two driven propellers that rotate in opposite directions. A special gearbox and shafting arrangement enables a single engine to drive the two propellers. This system enables recovery of some of the slipstream rotational energy that would normally be lost. One or two installations have been made on ocean-going ships and significant efficiency gains have been achieved.

9.11.5 CLT propeller

In the Contracted and Loaded Tip (CLT) propeller, the blade tips are fitted with pieces at right angles to the plane of rotation. The initial impression is that the blade edges have been bent over towards the face, ie away from the ship. The attachments at the blade tips serve to generate thrust across the whole propeller blade and thus improve the propeller efficiency. A nozzle surrounds the propeller and a tunnel structure under the stern on either side is used to direct the incoming flow of water.

9.11.6 Grim Wheel

The Grim Wheel, or vane wheel, is mounted aft of the main propeller and is larger in diameter. It is a freely rotating propeller with high aspect ratio blades which vary from a coarse pitch at the boss to a very fine pitch at the tip. The wheel is rotated by, and extracts energy from, the propeller slipstream and produces an additional thrust from the tip region of its blades.

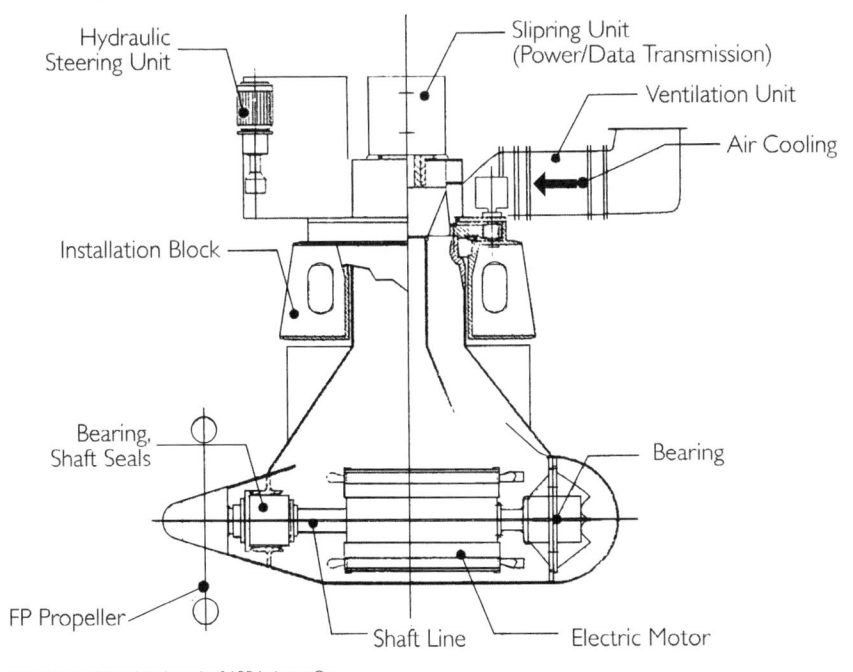

Azipod is a registered trademark of ABB Industry Oy

Fig 9.11: Azipod propulsion unit

9.11.7 Thrusters

A thruster is usually considered to be a device which assists in docking, manoeuvring, or positioning, of a vessel which is moving at a low speed. A fixed, or controllable pitch, propeller is used to move water either freely or in a duct. The complete unit may be fixed in position, eg a bow thruster, exposed or retractable, or able to rotate (azimuth).

The podded propulsor is a variation of the thruster which is being fitted to an increasing number of ships, particularly cruise liners. The podded propulsor may be fixed in position or able to azimuth, one example of the latter being the ABB Azipod. The Azipod is a podded electric propulsion unit, see Fig 9.11, capable of turning through 360deg. An unusual aspect of the device is that it normally pulls a vessel through the water, rather than pushing it like a conventional propulsion system. Located at the vessel's stern, Azipods eliminate the need for rudders, long lines of shafting and stern thrusters. They also offer improved manoeuvrability and considerable fuel efficiency.

9.11.8 Waterjet
In the waterjet system of propulsion, water is drawn into a duct beneath the vessel by an impeller within the duct which then accelerates the water and forces it aft, see Fig 9.12. The stator section of the duct contains guide vanes, which straighten out the rotation of the flow. The jet of water is then forced out through a nozzle, at a speed about twice that of the vessel.

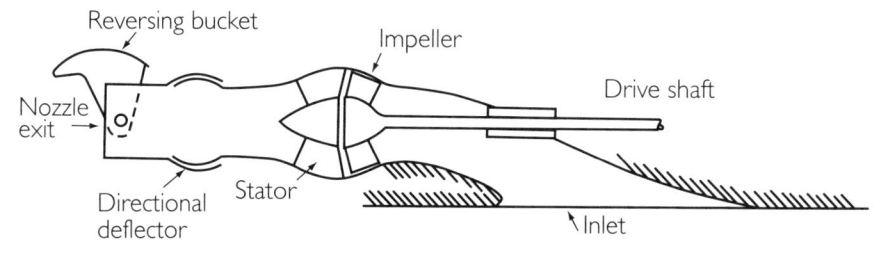

Fig 9.12: Waterjet

Steering, and the reversal of thrust for an astern movement, is brought about by directional deflectors that can be mechanically or hydraulically operated. The steering deflectors can be moved to 45deg port or starboard. The sideways force generated depends only upon the jet velocity and not on the velocity of water approaching, as with a conventional rudder. The considerable steering force thus produced enables the waterjet-powered vessel to turn within its own length.

Reversing is brought about by a deflector that moves down over the steering deflectors. In an intermediate position the deflector can equalise the water flows. This results in a neutral vertical thrust, which brings the boat to a stop with the impeller pump still operating. Water jet propulsion is being used on larger and larger ferries, since it is efficient, avoids cavitation problems, and enables reduced-draught operation.

9.11.9 Wind assistance
A number of attempts have been made to augment ships' propulsion systems by the use of sails, usually at times of high fuel costs. A folding sail arrangement has been used on one ship, and rotating cylinders using the Magnus effect have been considered.

10 Manoeuvring and motion control

The rudder is used to steer the ship. It is the principal control surface and directs the ship's movement in a horizontal plane. The different types of both conventional and specialist manoeuvring devices will be examined and the tests used to determine their effectiveness. Other control surfaces that influence a ship's movement in other planes will also be considered. These include stabilising systems and various types of foils which raise a ship vertically out of the water.

10.1 Directional control

A ship must be controllable with respect to direction in the horizontal plane. It may be necessary to maintain a particular heading or course, to turn to port or starboard, or even make some collision avoidance manoeuvre. The system used must be reliable and able to operate under all conditions the ship may encounter. The requirements for manoeuvring a ship include maintaining a specific course, obtaining a reasonable response when required, and being able to turn a complete circle of a reasonable diameter. In most vessels the rudder provides a vertical surface located aft to provide directional control. Some vessels requiring additional manoeuvrability may also have bow thrusters fitted. There are also various special steering devices and patented designs of rudders that will be considered separately.

10.1.1 Directional stability

A moving ship is considered to be directionally stable if, after being disturbed in a yawing motion, and with no corrective rudder action, it continues on some new path in a straight line. The disturbing force may arise from wave action in a disturbed sea. A ship requires some degree of directional stability but not too much, otherwise it would make the ship difficult to turn when a change of course was required.

An important feature in the directional control of a ship is the **neutral point**. This is the point in the length of the ship at which an applied force does not cause the ship to deviate from a constant course or direction. This neutral point is, in general, about one-sixth of the length of the ship aft of the bow. Thus, if a force is applied aft of the neutral point and acts towards port, the ship will turn to starboard. Again, if a force is applied forward of the neutral point and acts towards port, the ship will turn to port. Hence, the greater the distance the applied force is from the neutral point, the greater will be the turning effect. Thus rudders, which are used to bring about a direction-controlling force, are more effective when located aft.

10.1.2 The rudder

The rudder is, like the propeller blade, an aerofoil. It has a streamlined cross-section and will have lift (L) and drag (D) forces acting upon it when moved to an angle of incidence, α, relative to the water flow, see Fig 10.1. The resultant, F, of these two forces acts at the **centre of pressure** of the rudder, a position which varies with the angle of incidence. This resultant force can be determined from the expression:

$F = k\rho AV^2 f(\alpha)$,

where V is the water velocity past the rudder, A is the area of the rudder and ρ is the density of the water.

The coefficient, k, will have a value related to the cross-sectional and profile shapes of the rudder. The value of $f(\alpha)$ will increase almost linearly with the angle of incidence, until an angle between 35 and 45deg is reached. Stalling will then occur, due to a breakdown of flow around the rudder and $f(\alpha)$ will begin to reduce. Most ships' rudders are limited to 35deg maximum angle for this reason, and others that will be outlined later.

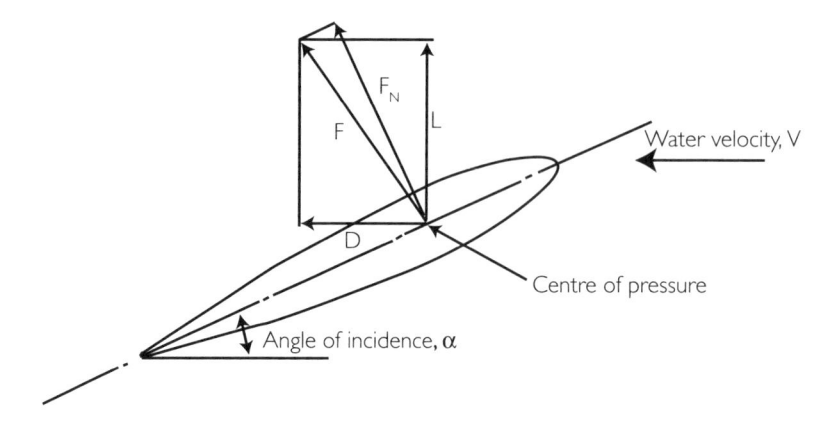

Fig 10.1: Forces on a moving rudder

This resultant force can be further resolved into forces normal to, and along, the longitudinal axis of the rudder. The axial force, which acts along the longitudinal axis, has no effect on the turning action and will not be considered further. The normal force, F_N, is regarded as the rudder force which acts to turn the ship.

The calculation of this force has been the subject of a number of formulae. When converted into the SI system of units, one of the older versions is:

rudder force = $577AV^2\sin(\alpha)$

where A is the rudder area and V is the velocity of water passing it.

To allow for the effect of the propeller race on the water velocity, the ship speed is increased by a factor of 1.3 for a rudder behind the propeller and 1.2 for a centreline rudder behind twin screws. A more recent formula for a centreline rudder behind a single propeller is:

rudder force = $18AV^2\alpha$

where V is the true speed of the ship and α is expressed in degrees.

The area of the rudder is usually decided upon the basis of experience with successful designs. This has resulted in rudder area being some percentage of the product of the ship's length and its draught.

Typical merchant ship values are in the range of 1.4 to 1.6% or, when expressed as a ratio of rudder area to the product of length and draught, 1:70 to 1:60. Only one classification society, Det norske Veritas, actually specifies a minimum rudder area. For a rudder working directly behind a propeller, the minimum area is given by the expression:

$$\text{rudder area, A} = \frac{Ld}{100}\left(1 + 25\left(\frac{B}{L}\right)^2\right)$$

where L is the ship length, d is the draught and B is the breadth.

The shape of a rudder is often dictated by the stern arrangement of the vessel. The most important shape parameter is the aspect ratio, which is the mean span or depth, divided by the mean chord or width. Most rudders on merchant ships have an aspect ratio less than two. A high aspect ratio rudder would give a higher lift coefficient for a given angle of incidence, but would also stall at a lower angle of incidence.

All rudders are made up of the control surface, or blade, and the vertical rudder stock, which transfers the rotary motion of the steering gear to the blade. There are three general types of rudder. These are; unbalanced, semi-balanced and balanced.

An **unbalanced rudder** has all of its blade area located aft of the rudder stock. The **semi-balanced rudder** has a portion of its area, less than the full height or span, forward of the rudder stock. The **balanced rudder** has a full span portion of its area located forward of the rudder stock.

The term 'balanced' relates to the torque required to move the rudder in relation to the rudder force. The rudder force acts at the centre of pressure of the rudder and this position varies with the angle of incidence. The centre of pressure moves aft towards the trailing edge of the rudder as the angle is increased. A balanced rudder is designed to have the centre of pressure coincident with the centreline of the rudder stock at an angle of incidence of about 15deg. In this balanced condition there is a zero torque on the rudder and the steering gear is unloaded. The power requirements for a steering gear operating a balanced rudder will, therefore, be less than for an unbalanced type.

The centre of pressure, or centre of effort, for a rudder will vary with the shape of the rudder. It will also change position and move aft as the angle of incidence increases. An estimate of this centre of pressure position is necessary when calculating the torque acting on a rudder, with a view to determining steering gear and rudder stock strength requirements. For a rectangular flat plate, one empirical formula to determine the position of the centre of pressure is:

centre of pressure aft of leading edge = $b(0.195 + 0.305\sin\alpha)$
where b is the width and α is the angle of incidence.

The centre of pressure will vary between 0.195b and 0.37b as the rudder moves from midships to 35deg. More recent work, based on aerodynamic data, considers chord width to allow for non-rectangular rudders. It suggests a variation in centre of pressure position of 0.15b to 0.27b for rudder movements from midships to 35deg.

Strength calculations for rudders usually consider only the rudder force which acts normal to the rudder, and the torque that is required to turn the rudder. The most severe operating conditions are usually considered and then allowances are made for the impact of heavy seas. The calculation of the rudder force and also the position of the centre of pressure was outlined earlier. If the distance of the centre of pressure from the centreline of the rudder stock – ie the moment arm – is found, then the torque on the rudder stock can be calculated. Hence:

torque on rudder stock, T = rudder force × moment arm
$577AV^2\sin(\alpha)$ × moment arm
(or $18AV^2\alpha$ × moment arm)

The basic torque equation may then be used to determine the diameter of the rudder stock. Thus:

$$\frac{T}{J} = \frac{q}{r}$$

where J is the second moment of area about a polar axis, r is the radius of the rudder stock, and q is the allowable stress in the metal.

Example

A rudder with an area of 20m² when turned to 35deg has the centre of effort 1.2m from the centreline of the stock. If the ship speed is 15 knots, and the rudder is located aft of the single propeller, calculate the diameter of the rudder stock. The allowable stress is 70MN/m².

Rudder force = $577AV^2\sin(\alpha)$
= 577 x 20 x (15 x 1.3 x 0.5144)² x sin35deg
= 0.666MN

Note: The factor 1.3 is used to allow for propeller race effect.

Torque on rudder stock = Rudder force x moment arm
= 0.666 x 1.2
= 0.7992MNm

Since $\dfrac{T}{J} = \dfrac{q}{r}$ and $J = \dfrac{\pi r^4}{2}$

then $r^3 = \dfrac{T \times 2}{\pi \times q} = \dfrac{0.7992 \times 2}{\pi \times 70}$
= 0.00727

∴ r = 0.194m and diameter of rudder stock = 0.388m.

If the alternative equation, (rudder force = $18AV^2\alpha$), had been used, a diameter of 0.416m would be required.

10.1.3 Rudder operation and ship turning

When a rudder is moved to some angle, a rudder force is generated which acts at the centre of pressure and normal to the rudder surface. The small force which acts along the surface of the rudder has no effect on turning and will not be considered further. The rudder force, F, results in a moment, FA, about a vertical axis through the ship's centre of gravity, G, see Fig 10.2(a). The horizontal plane considered will be through the centre of pressure of the rudder. The point B is where this plane intersects the vertical axis through the ship's centre of gravity.

For the ship to turn, a force must act which is directed towards the centre of the turn. The rudder creates a moment acting on the ship and maintains an angle of attack in relation to the water flowing past the ship.

The force, F, may be replaced by an equivalent force, F_1, and a couple, FA, both acting at B. The force, F_1, has two components $F_1\cos\theta$ and $F_1\sin\theta$, which are vertical to, and act along, the ship's centreline respectively. The motion of a ship, when turning, can be considered as three separate phases during which the forces acting will vary considerably.

The movement of the rudder begins the first phase, see Fig 10.2(a). The ship will accelerate to port as a result of the force $F_1\cos\theta$. The force $F_1\sin\theta$ will oppose the ship's forward motion, and it will slow down. The couple FA will cause the ship to rotate about the axis through B in the desired direction, ie to starboard. As the rotational inertia of the ship is overcome it will move through some angle α.

The second phase of the turn now begins as the instantaneous velocity of the ship acts at an angle α to the centreline, see Fig 10.2(b). The ship itself now acts as an aero-

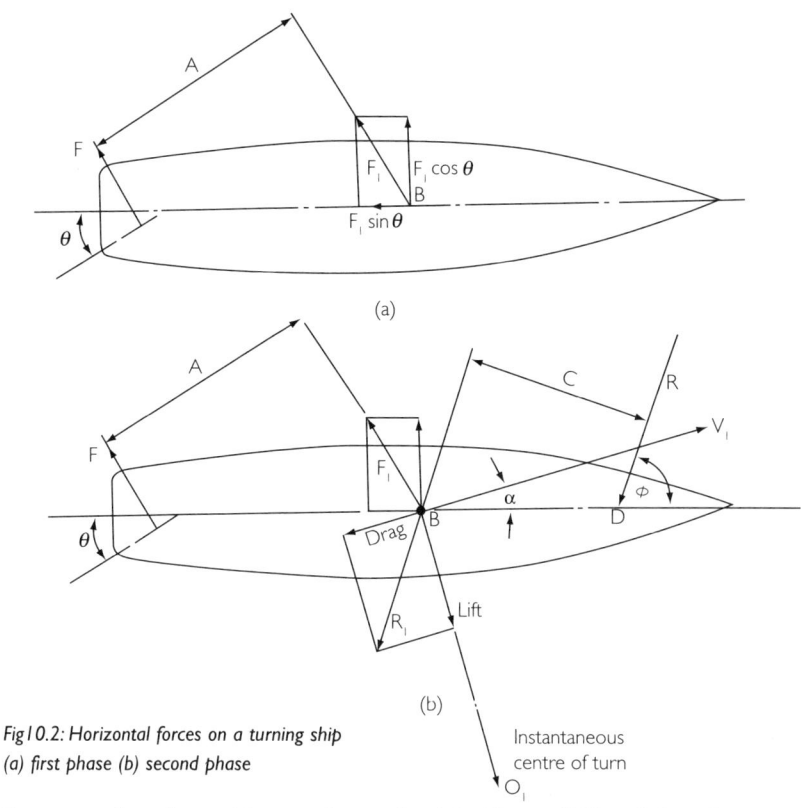

Merchant Ship Naval Architecture

Fig 10.2: Horizontal forces on a turning ship
(a) first phase (b) second phase

foil or control surface with α as the angle of incidence. Lift and drag forces create a resultant force R, which acts at some point D on the ship's centreline, at an angle ϕ. This force R may be replaced by an equivalent force, R_1, and a couple, RC, both acting at B. There are now two couples, FA and RC, acting to turn the ship, and the angle of incidence, α, will increase. The point D will move aft as α increases and the couple RC will reduce. This creates something of an s-shape in the first quarter of the turning circle. The outboard (to port) acceleration of the ship will be gradually overcome by the lift components of the force, R_1. The ship will ultimately be accelerated inwards (to starboard) towards the instantaneous centre of the turn. The drag components of R_1 will be further slowing the ship down.

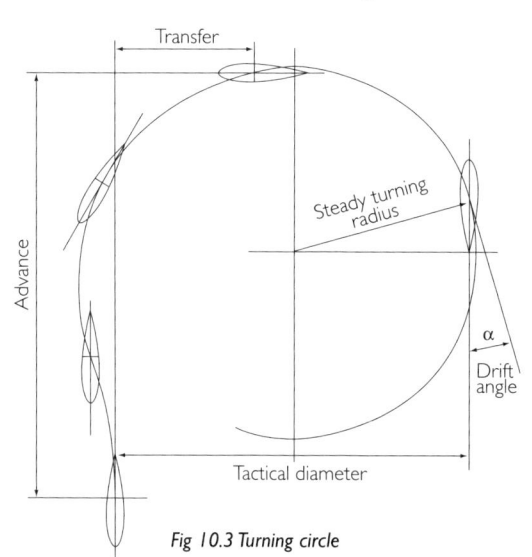

When the point D reaches a position aft of B, such that the couples FA and CD (now opposing FA) are equal, the ship enters the final or steady turning phase. The forces F_1 and R_1 now

Fig 10.3 Turning circle

have constant values and the ship is both slowed down and accelerated towards the, now fixed, centre of the turning circle. The inward acceleration is a constant value.

The complete motion path of the ship, when turning, is shown in Fig 10.3. Advance is the forward distance travelled by the centre of gravity of the ship in the first quadrant of a turn. Transfer is the athwartships distance travelled by the ship in completing the first quadrant of a turn.

In addition to turning when the rudder is moved, a ship will also heel. This heeling action results from the vertical disposition of the various forces. These forces will further vary, depending upon the phase of the turn, as discussed above.

The initial movement of the rudder begins the first phase. The transverse component of the rudder force, $F\cos\theta$, acts at the centre of pressure of the rudder, P, and away from the centre of the turning circle, see Fig 10.4 (a). The heeling axis of the ship will be at some position G, which is actually considered as the centre of gravity of the ship. A moment of $F\cos\theta \times PG$ will, therefore, tend to heel the ship towards the centre of the turning circle.

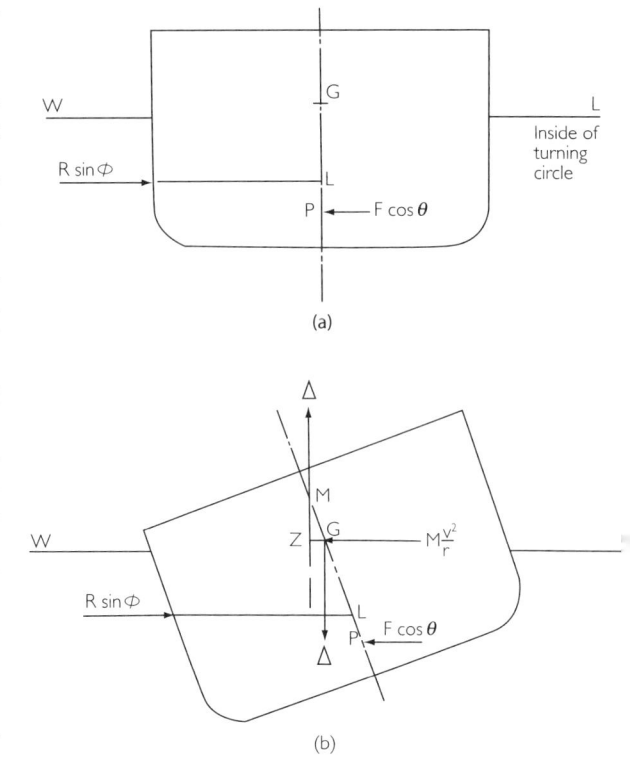

Fig 10.4: Heeling forces on a turning ship (a) initial phase (b) final or steady turning phase

The resultant force, R, develops during the second phase and creates a moment, $R\sin\phi$. This component is considered to act at the centre of lateral resistance, which is usually considered as half of the draught. This force component creates a moment, $R\sin\phi \times GL$, which opposes the heeling moment created by the component of the rudder force. As the resultant force component, $R\sin\phi$, increases, the ship will reduce its angle of heel towards the centre of the turning circle, become upright, and then heel away from the centre of the turning circle.

In the final, or steady turning phase, the ship will reach some angle of heel, θ, which results in equilibrium between the various forces acting, see Fig 10.4(b). When turning steadily, a centrifugal force resulting from the ship's mass will act at G:

$$\text{centrifugal force} = \frac{MV^2}{r}$$

where M is the ship's mass, V is the ship's speed and r is the radius of the turning circle. The ship will be in equilibrium when the three heeling moments are equal to the righting moment. If the angle of heel due to turning is likely to be small, then it can be determined by using the metacentric height, GM, to determine the righting

Merchant Ship Naval Architecture

moment. Thus:

heeling moment $= \dfrac{MV^2}{r}(LG\cos\theta) - F\cos\theta(PG\cos\theta)$,

righting moment $= MgGM\sin\theta$.

The heeling moment created by the force on the rudder is usually small and in some circumstances, may be ignored. This provides a simplified expression to determine the angle of heel when turning. Since,

righting moment = heeling moment

$$MgGM\sin\theta = \dfrac{MV^2}{r}(LG\cos\theta)$$

$$\dfrac{\sin\theta}{\cos\theta} = \dfrac{MV^2}{r} \dfrac{LG}{Mg \times GM}$$

$$\tan\theta = \dfrac{V^2 LG}{g \times r \times GM}$$

where θ is the angle of heel. The angle of heel is an approximate value and relates only to small angles, since it uses formulae related to small angle stability.

The force on the rudder can have a significant effect on the angle of heel when turning, if the rudder were suddenly turned in the opposite direction. Such an action would increase the angle of heel when it might, inadvertently, be thought to be reducing it.

10.1.4 Assessing manoeuvrability
A number of specific manoeuvres are used to assess the directional stability of a ship and its responsiveness to changes in rudder angle. The tests are usually carried out on sea trials.

Turning circle
The turning circle manoeuvring trial is conducted in calm seas with little or no wind. The vessel will be travelling at a steady speed on a straight course when a rudder deflection is applied and the vessel movement through 360deg is plotted.

Measurements are then made of the tactical diameter, and the advance and transfer distances following a 90deg change of heading or direction, see Fig 10.3. The time taken to change the heading by 360deg is also measured, as are the drift angle and angle of heel. The changes in ship speed and propeller shaft revolutions during the manoeuvre are also noted. The trial is usually repeated at different speeds and applied rudder angles.

Spiral manoeuvre
This trial determines whether or not a ship has directional stability. The ship's rudder is first turned through 15deg to starboard. When a steady turning rate is achieved, the value is recorded. The procedure is then repeated for starboard rudder angles of 10deg, 5deg, 4deg, 2deg, 1deg and 0deg. The rudder angle is then put through the same values to port, up to 15deg. The procedure is then reversed to finally return to a rudder angle of 15deg to starboard. The operation is carried out at a steady propeller shaft speed and is usually repeated at different shaft speeds.

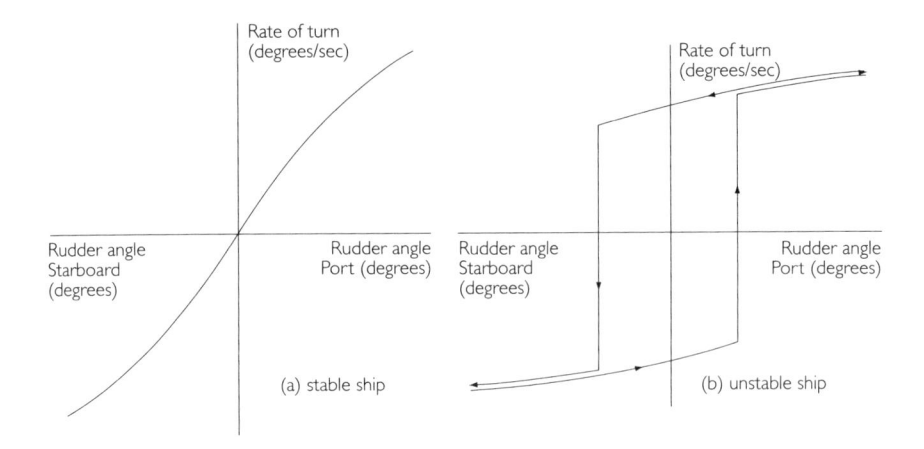

Fig 10.5: Spiral manoeuvre (a) stable ship (b) unstable ship

The plot of results should appear as in Fig 10.5(a) for a stable ship. A plot in the form of Fig 10.5(b) would indicate a ship that was unstable with respect to directional stability.

A directionally-stable ship will have a particular rate of change of heading at each rudder angle. A directionally-unstable ship will only have effective changes of heading when the rudder angle is large. The behaviour at small angles of rudder is difficult to predict, but can be shown with captive model tests to pass through the origin, as indicated in Fig 10.5(b).

Zig-zag manoeuvre

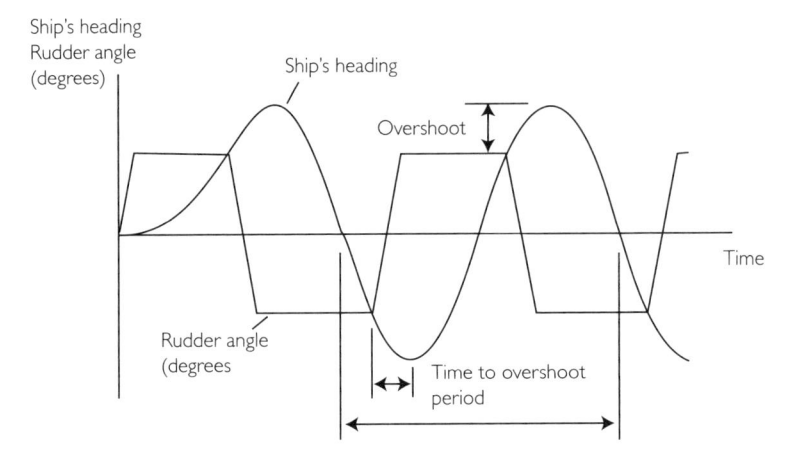

Fig 10.6: Zig-zag manoeuvre

The zig-zag manoeuvre enables an assessment of the ship's initial response to sudden rudder movements. When the ship is proceeding at a steady speed on a straight line course the rudder is quickly and smoothly positioned at 20deg to starboard and held there until the ship changes heading by 20deg. The rudder angle is then quickly and smoothly moved to 20deg to port and held until the ship changes heading by 20deg. The process is continued and a plot is produced of the ship's course against rudder angle as shown in Fig 10.6. Measurements are made of overshoot, period, and time to overshoot, for comparison with an acceptable ship's movements. The operation can be repeated for different ship speeds and different amounts of rudder angle.

10.2 Types of rudder

Streamlined rudders of a conventional double-plate construction are fitted to almost all modern merchant ships. Ferries and other vessels with particular ship-handling requirements often make use of some specialist rudder types and manoeuvring devices.

10.2.1 Conventional

Conventional rudders can be further described by their arrangement about the turning axis as unbalanced, balanced or semi-balanced. This was discussed earlier with reference to torque required to turn the rudder. The after end arrangement of a vessel will also, to some extent, determine the type of rudder fitted.

A balanced rudder is shown in Fig 10.7(a). The centre of pressure will be coincident with the centreline of the rudder stock at an angle of incidence of about 15deg. The rudder has an axle fitted at its turning axis and rotates about bearings fitted top and bottom. The rudder stock is fitted to a palm aft of the axle, but is cranked so that it finally rises coaxially with the axle. Another form of balanced rudder is the spade type, see Fig 10.7(b). The rudder stock acts as the axle in this arrangement.

An unbalanced rudder is shown in Fig 10.8, where it can be seen that no part of the rudder area is forward of the turning axis. The rudder turns about pintles fitted top and bottom. The bearing housings, or gudgeons, form part of the streamlined sternpost.

The semi-balanced rudder shown in Fig 10.9 is of the spade type. A portion of the rudder area can be seen to project forward of the turning axis. The rudder turns about pintles, which have bearing housings on the horn of the sternframe.

10.2.2 Special rudders and manoeuvring devices

The rudder is basically a manoeuvring device. Various special designs of rudder exist, some of which are patented. In certain cases special propulsion arrangements

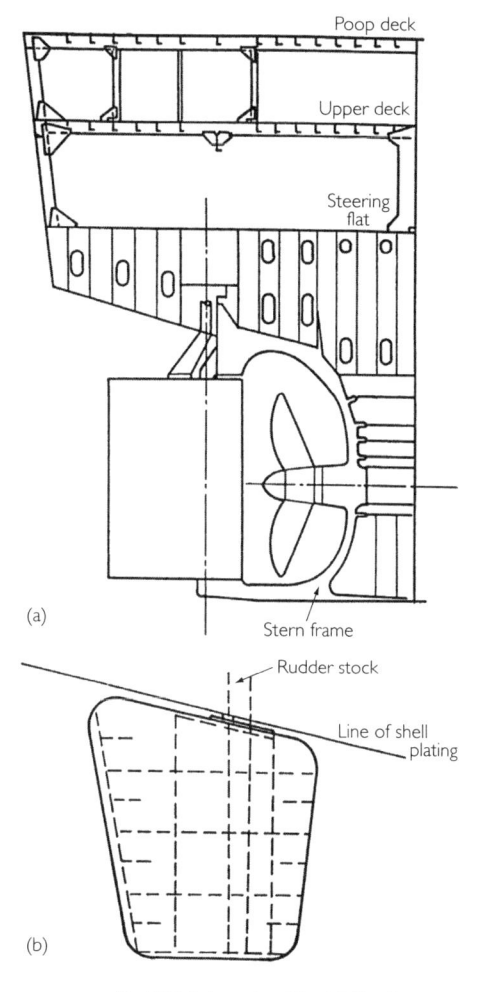

Fig 10.7: Balanced rudder (a) Simplex type (b) spade type

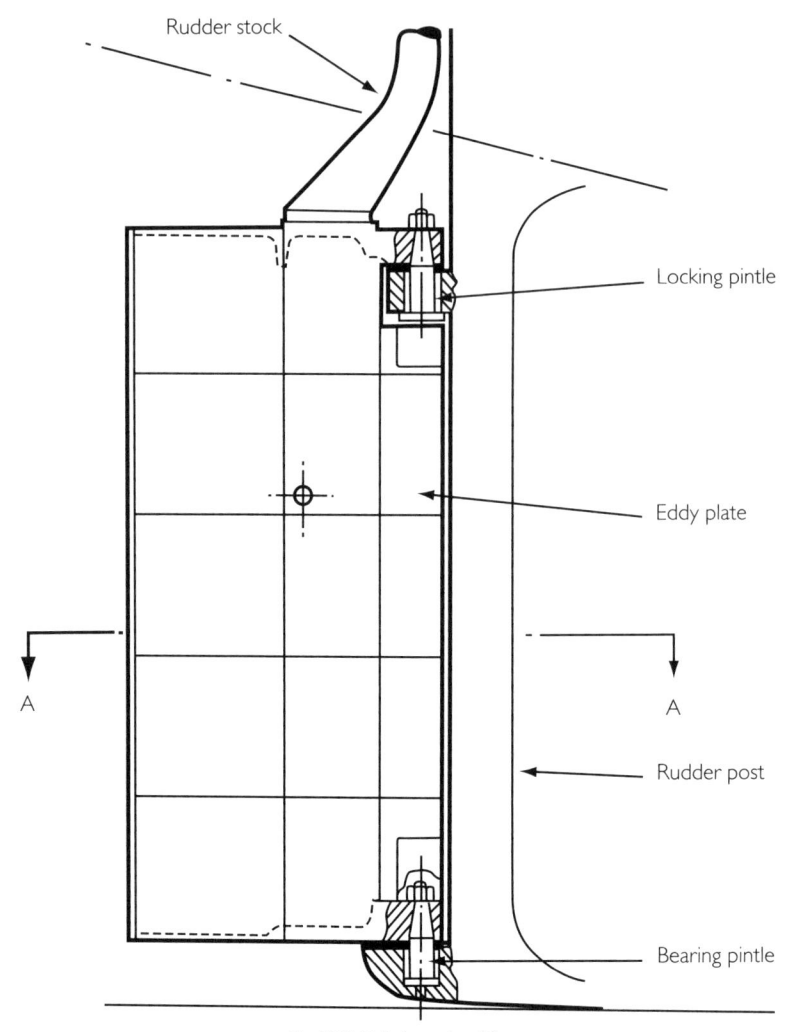

Fig 10.8: Unbalanced rudder

enable manoeuvrability without an actual rudder. A number of these special devices will now be examined.

The flap rudder makes use of a narrow flap at the trailing edge of the rudder. The flap is operated at low or moderate ship speed and moves to a greater angle of incidence than the main rudder. The aerofoil shape of the rudder is, therefore, changed to give a greater lift force. At high ship speed, when the flap is not required, it is aligned with the main body of the rudder. It is possible to position the main rudder fore and aft and just use the flap movements for all steering.

The Flettner rudder is a special type of flap rudder which uses two narrow flaps at the trailing edge. The flaps move in such a way that they assist the movement of the main rudder. The reduced forces needed to turn the Flettner rudder will result in the use of a smaller steering gear than for a conventional rudder.

Where a rudder arrangement results in a streamlined fixed structure followed by a movable rudder, the two can be considered as a flap rudder. A greater lift force results, with less drag and less rudder torque needed. The fixed structure may be a

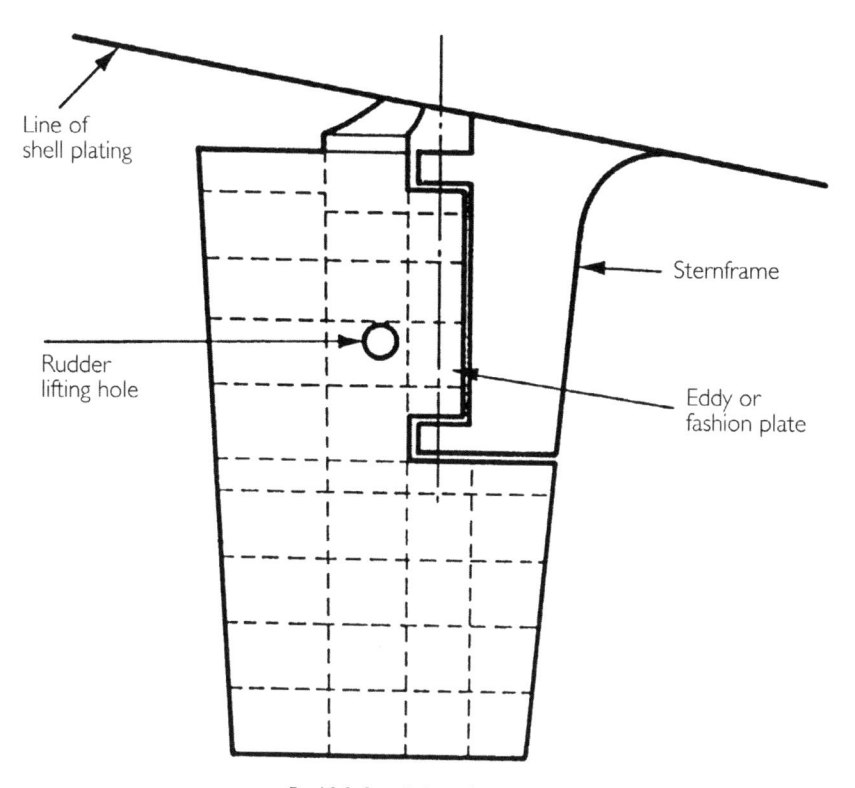

Fig 10.9: Semi-balanced rudder

sternpost, a horn or the skeg of the ship. A particular design of this type is the Oertz rudder.

A rotating cylinder placed at the leading edge of a conventional or flap rudder will further increase the lift force. The high-speed rotation of a cylinder in a fluid stream develops a lift normal to its axis and the stream flow. This is called the Magnus effect. The rotating cylinder can be used alone as a rudder. The combination arrangement enables rudder angles of up to 90deg without stall, and a considerable reduction of the ship's turning-circle diameter.

The Kitchen rudder, in addition to steering the ship, regulates the speed and enables astern propulsion. The rudder is a two-part tube, which shrouds the propeller. It is hinged about a vertical central axis such that the tube may move or the two halves may be moved, see Fig 10.10. The deflection of the propeller race (water flow) provides the steering action and variation of the thrust direction.

Bow rudders have been fitted to a few ships, but are much less effective than the conventional stern rudder, for reasons outlined earlier. Use is usually made of a bow thrust unit to improve manoeuvring of a vessel when berthing. A bow thruster is a propeller within a transverse tube, which can provide thrust to port or starboard, as required.

10.3 Motion control

A number of other forms of control surfaces will now be considered which deal with ship motions. Rolling is the only ship motion that can be effectively dampened and reduced. A number of control surfaces can be used, independently or together, these

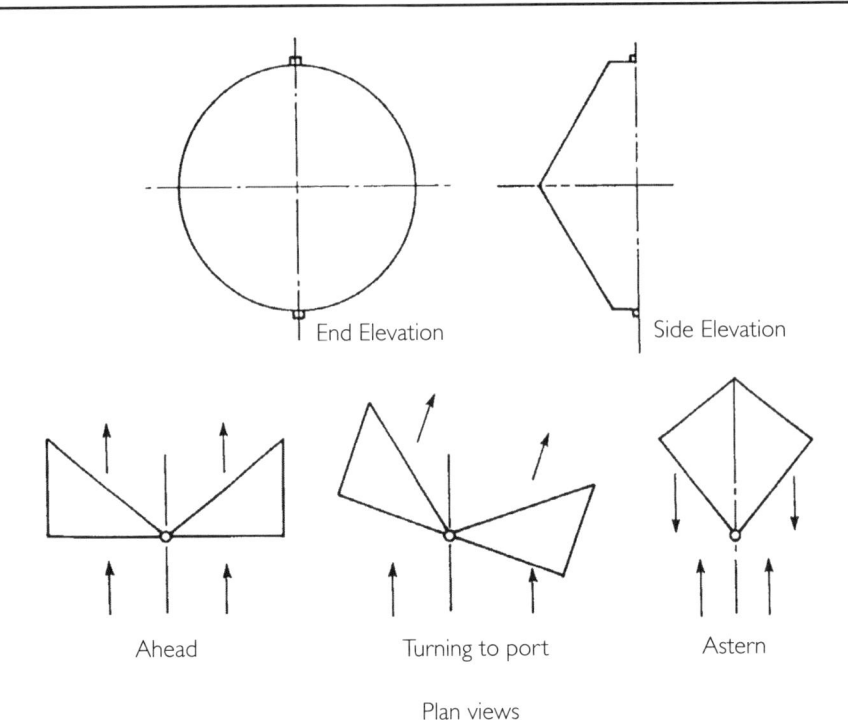

Fig 10.10: Kitchen rudder

being; bilge keels, fin stabilisers and tank stabilisers.

The rolling of a ship is probably the most objectionable motion from the points of view of human comfort and cargo damage. Various methods, both passive and active, can be used to reduce rolling, and consideration will now be given to bilge keels, stabilising fins and stabilising tanks. All three can be considered to use control surfaces to bring about their operating action.

10.3.1 Bilge keels

A bilge keel is a passive device comprising a permanent flat plate projection from the ship's bilge region which extends for the midship section length. They are fitted at right-angles to the radiused bilge-plating on each side of the ship, and project a sufficient distance to penetrate the boundary layer of water along the hull.

The operation of a bilge keel is based upon the equation for the rolling period:

$$T_R = 2\pi \frac{K}{\sqrt{gGM}}$$

The value of the mass radius of gyration, K, will be increased, since the projecting keels will force the movement of more water. The rolling period will, therefore, be increased and the amplitude of roll is decreased.

The bilge keel also creates a damping moment resulting from the pressure resistance of the projecting area and the viscous eddy flows around it. Bilge keels have been found to be more effective when a ship is moving. This seems to imply some hydrodynamic effects, possibly related to the aerofoil section theory considered in relation

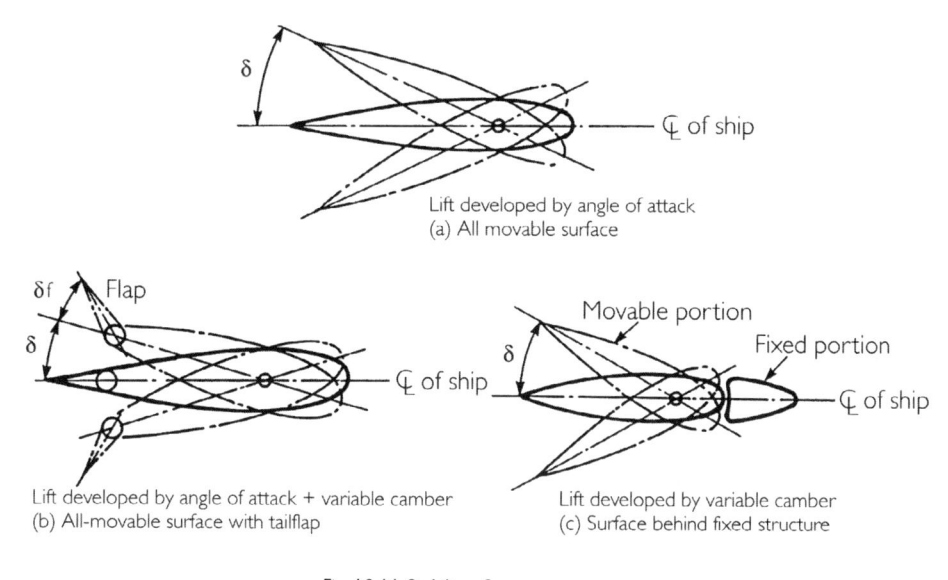

Fig: 10.11 Stabiliser fin arrangements

to propellers and rudders. The bilge keel does, however, provide some roll reducing action even with a stationary ship. Almost all ships are fitted with bilge keels.

10.3.2 Fin stabilisers

Fin stabilisers are active devices which make use of one or more pairs of fins which are fitted, one on each side of the ship. The size or area of the fins is governed by ship factors such as breadth, draught and displacement, but is small compared with the size of the ship. The fins may be retractable, ie pivoting or sliding within the ship's form, or permanently extended. They act to apply a righting moment to the ship as it is inclined by a wave, or force, on one side. The angle of incidence of the fin, and the resulting moment on the ship, is determined by a sensing control system.

The stabilising force is generated by the lift on fins of aerofoil section. The fins may be of the all-moveable type without flaps, see Fig 10.11(a) or with flaps as in Fig 10.11(b). They may also be partly fixed and partly movable, as shown in Fig 10.11(c). The angle of incidence of the fin is controlled in order to produce an upward force on the fins on one side of the vessel and a downward force on the other. The resulting fin moment will then oppose the rolling moment.

The lift force can be found from the expression:

$$\text{lift force} = k\rho AV^2 f(\alpha)$$

where V is the velocity of the ship, A is the fin area and ρ is the density of sea water.

The coefficient k will have a value related to the cross-sectional and profile shapes of the fin. It can be seen that when the ship speed is zero, there will be no stabilising force. Also the stabilising force will increase by the square of the increase in speed. Where a ship has a wide range of operating speeds, the fin angle, α, can be suitably modified to provide a fairly constant stabilising ability over that range.

A typical folding stabiliser fin arrangement is shown in Fig 10.12. The fin movement is brought about by a hydraulic power unit. This is controlled by a system that utilises an angular accelerometer which continuously senses the rolling accelerations

Fig: 10.12 Fin stabiliser

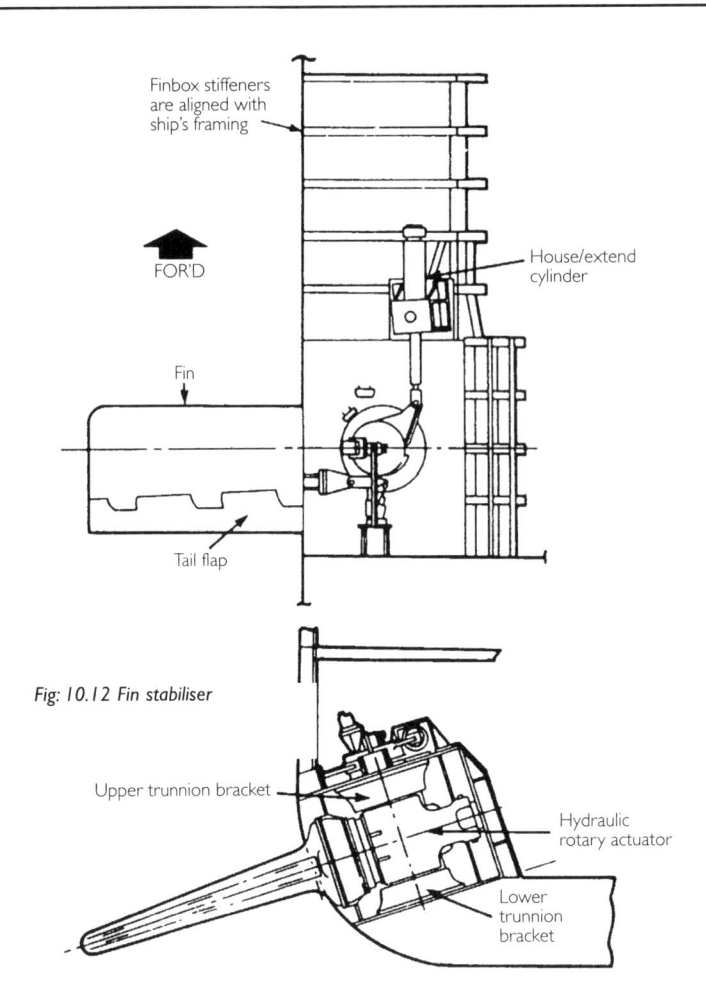

of the ship. As a result of the control system, fin movement is a function of roll angle, roll velocity, roll acceleration and natural list. Fin stabilisers provide accurate and effective roll stabilisation but are complex installations.

10.3.3 Tank stabilisers

A tank stabiliser provides a righting or anti-rolling force as a result of the delayed flow of fluid in a suitably positioned transverse tank. The system operation is independent of ship speed and will work even when the ship is at rest.

Consider a mass of water in an athwartships tank. As the ship rolls, the water will be moved, but at a very short period of time after the ship. Thus, when the ship is finishing its roll and about to return, the still moving water will oppose the roll. The water mass thus acts against the roll at each ship movement. This athwartships tank is sometimes referred to as a 'flume'. The system is considered passive, since the water flow is activated by gravity.

A wing tank system arranged for controlled passive operation is shown in Fig 10.13. The greater height of tank at the sides permits large water build-up and, thus, a greater moment to resist the roll, but the rising fluid level must not, however, fill the wing tank. The air duct between the two wing tanks contains valves that are operated

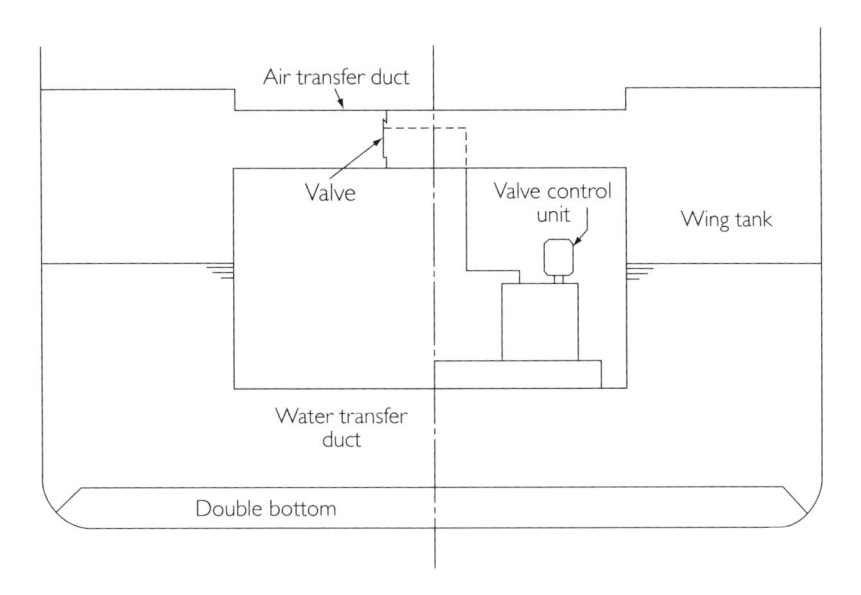

Fig: 10.13 Air-controlled tank stabiliser

by a roll-sensing device. The differential air pressure between tanks is regulated to allow the fluid flow to be controlled and 'phased' for maximum roll stabilisation.

A tank system must be specifically designed for a particular ship, by using data from model tests. The water level in the system is critical, and must be adjusted according to the ship's loaded condition. Also, with a tank system, there is a free surface effect resulting from the moving water, which consequently reduces the stability of the ship. The tank system does, however, stabilise a ship which is not under way, and is a much less complex installation than a fin stabiliser.

10.4 Vertical lift control

The vertical lift control considered here, refers to the use of foils to raise the vessel out of the water, as in hydrofoil craft. Reduced resistance and improvements in ship motion result with this type of craft.

10.4.1 Hydrofoils

The lift force produced by a hydrofoil moving through the water can be used to support the vessel clear of the water on struts. This foil, or control surface, will develop a lifting force that increases with the square of the speed. If the propulsion unit is sufficiently powerful, the vessel will eventually lift from the water. The hull resistance will then cease to exist and further speed increase is possible, until the resistance of the foils, struts, appendages and the air absorb the available power.

Earlier work in this chapter it was shown that the lift force can be found from the expression:

lift force = $k\rho AV^2 f(\alpha)$,

where V is the velocity of the ship, A is the foil area and ρ is the density of the water.

The coefficient, k, will have a value related to the cross-sectional and profile shapes of the foil and α is the angle of attack.

When the hull is raised up clear of the water, the lift force required from the foils will be constant. If the speed is now increased, then the angle of incidence, or the immersed area of the foil must reduce, in order to keep the lift force constant. Two basically different types of foil have been developed to meet these requirements.

Surface piercing foils have reducing immersed area as the craft increases speed through the water. This is because the increasing lift force causes the vessel to lift higher from the water. Completely-submerged foils can vary their angle of incidence, in order to change the lift force, or maintain it constant, with increasing speed of the craft.

The surface piercing foil arrangement may use large dihedral V-shaped foils or multiple small V-shaped foils in tiers on either side of the craft. Surface piercing foils maintain the craft upright and correctly trimmed, without the need for a control system. The foils effectively sense and provide any required restoring forces and moments to counter the effect of change in speed or wave action. However, the foil area providing this self-stabilising action is subject to wave action. The craft will, therefore, be subject to motion in a seaway. In order to provide longitudinal balance, a large foil area is provided either forward or aft of the centre of gravity and a smaller foil is located at the other end of the craft.

Completely-submerged foils adjust to changing lift requirements by a change in the angle of incidence of the complete foil or its attached flaps. A sensing and automatic control system is required to maintain the craft at a particular height above the water. This control system, which operates on the foil flaps both port and starboard and fore and aft, will keep the craft upright and correctly trimmed at some particular height from the water surface. Since the foils are fully submerged, this arrangement is subject to fewer wave-induced motions than the surface piercing type. Longitudinal balance is provided by almost equal foil lengths located fore and aft of the craft.

The surface piercing foil craft will follow the contours of ahead waves and thus will pitch to some extent. It is unlikely, however, that the hull of the craft will strike the waves at 'bottom'. The completely submerged foil craft relies upon its automatic control system to adjust to oncoming waves, and will be largely unaffected in its forward motion. The surfacing piercing foil craft is unable to cope with following seas, whereas the fully submerged type can operate satisfactorily.

11 Vibration

Ship vibration is a subject of considerable importance to naval architects and marine engineers. For decades, the subject has demanded the attention of ship technologists. It is a difficult problem to tackle as the ship is a complex structure and there are many causes of vibration. The adverse effects of vibration can cause fatigue damage and, in serious cases, even structural failure. Shipboard equipment too may be damaged or systems may malfunction. Passengers and ships' personnel may find vibration a source of discomfort, which could affect their work performance. Most machines and structures experience vibration to some degree. Although ship vibrations cannot be entirely eliminated, they must be reduced as far as possible. In this chapter, the general characteristics of a vibrating system will be considered first, before proceeding to the subject of ship vibration.

11.1 Basic concepts

The simplest vibration model is an oscillatory system that consists of a mass, a spring with negligible mass which produces a force proportional to the displacement and a damper which produces a force proportional to the velocity. Neglecting damping, the resulting spring-mass system, see Fig 11.1, can be shown to undergo simple harmonic motions when it moves in a vertical direction. The equation of motion is:

$$m\left(\frac{d^2x}{dt^2}\right) + kx = 0$$

putting $\omega_n = \sqrt{(k/m)}$, where k being the spring constant,

$$\frac{d^2x}{dt^2} + \omega_n^2 x = 0$$

The natural frequency of the system, f_n, is equal to $\omega_n/2\pi$. The natural frequency, f_n, is an important property of the system. This model can be applied to other systems with single degree of freedom, such as a beam undergoing lateral or torsional vibrations under various end conditions, eg simply-supported or with clamped ends.

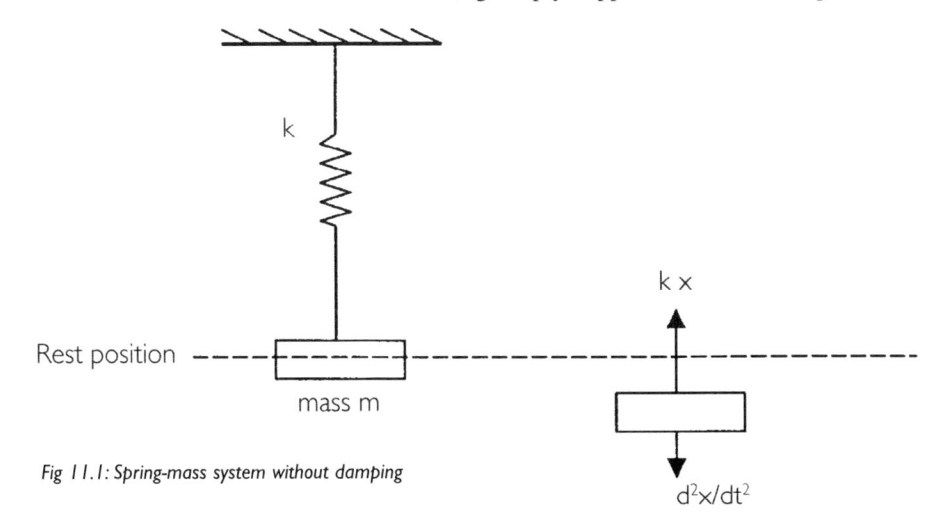

Fig 11.1: Spring-mass system without damping

Without damping, the system could, in theory, oscillate indefinitely. However, in reality, damping forces, such as air resistance, exist and cause the amplitude of motion to die down with time.

11.1.1 Damping effects

To realistically model vibrating systems, damping cannot be ignored. The simplest damping form is one that varies linearly with velocity of motion ie c dx/dt, where c is a damping coefficient. A system with damping is represented by a dashpot, see Fig 11.2. The equation of motion is:

$$m\left(\frac{d^2x}{dt^2}\right) + c\left(\frac{dx}{dt}\right) + kx = 0$$

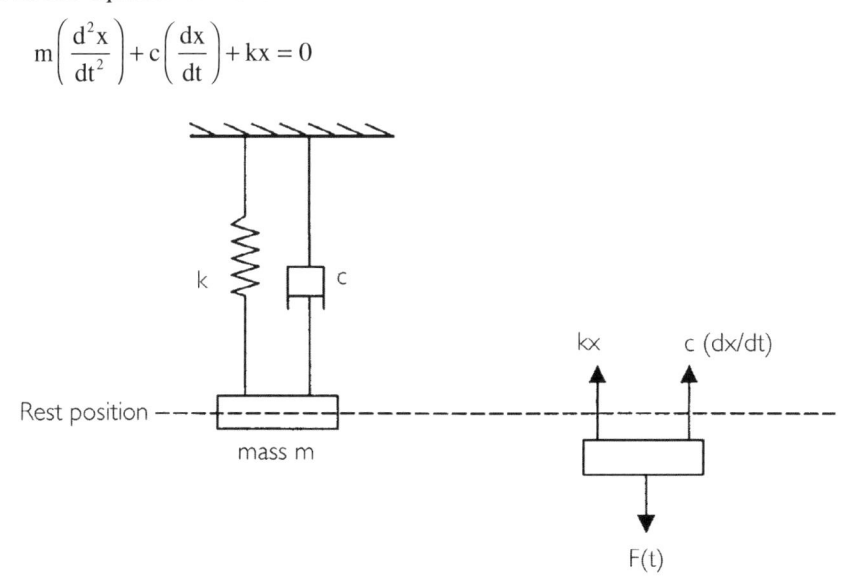

Fig 11.2: Spring-mass system with liner damping

For many systems, damping would be light. The resulting motions would be a series of complete oscillations of steadily diminishing amplitudes, which eventually comes to rest. A general solution would be:

$$x = A \exp(-\alpha t)\sin(\beta t + \delta)$$

where A is the initial amplitude of motion, δ the phase angle and α, β are constants to be determined. It can be shown for this type of motion, the solution is:

$$x = A\exp\left(-\left(\frac{c}{2m}\right)t\right)\sin\left\{t\sqrt{\left[\left(\frac{k}{m}\right) - \left(\frac{c}{4m^2}\right)\right]} + \delta\right\}$$

The rate of motion decrement depends on (c/2m), that is when damping is large relative to the mass of the system, motion would die down quickly. Damping also has an influence on the natural frequency which is given by $\{(k/m)-(c/4m^2)\}^{0.5}/2\pi$. For many practical applications, c is likely to be small and the natural frequency is often taken as the natural frequency of free vibration $(k/m)^{0.5}/2\pi$. The system is called critically damped when $(k/m)=(c/4m^2)$, and the resulting motion is non-oscillatory.

11.1.2 Forced vibration

In the preceding section, the system was displaced from its equilibrium position and allowed to oscillate freely. This is known as free vibration. In practice, external forces may be applied continuously over a long period and vary in magnitude. In this case, the disturbing force can be written:

$$F = F_0 \sin(\omega t)$$

where F is a sinusoidal force with amplitude F_0 and a forcing frequency of $\omega/2\pi$.

The equation of motion for the vibrating system becomes

$$m\left(\frac{d^2x}{dt^2}\right) + c\left(\frac{dx}{dt}\right) + kx = F_0\sin(\omega t)$$

and the resulting oscillation after a steady state is reached, is of the form

$$x = x_0 \sin(\omega t + \delta)$$

where x_0 is the amplitude of oscillation and δ is the phase angle. Substituting the corresponding velocity and acceleration terms into the equation of motion, the in-phase and out-of-phase components are derived. From these components, expressions for the motion amplitude and phase angle would be:

$$\frac{x_0}{F_0} = \frac{1}{\sqrt{\{(k-m\omega^2)^2 + c^2\omega^2\}}}$$

$$\tan \delta = \frac{(-c\omega)}{(k-m\omega)}$$

Introducing a tuning factor $\Lambda = \omega/(\sqrt{(k/m)})$, which is the ratio of the forcing frequency to the natural frequency of the system, the expressions become:

$$\frac{x_0}{F_0} = \frac{1}{\left[k\sqrt{\left\{(1-\Lambda^2) + \left(\frac{c^2\Lambda^2}{mk}\right)\right\}}\right]}$$

$$\tan \delta = \frac{\left\{\frac{-c\Lambda}{(mk)^{0.5}}\right\}}{(1-\Lambda^2)}$$

The ratio, F_0/k, represents the displacement under a static force of F_0. The dynamic magnification factor is the ratio of dynamic and static displacements, ie x_0/X, thus:

dynamic magnification factor,

$$\frac{x_0}{X} = \frac{1}{\left[\sqrt{\{(1-\Lambda^2)^2 + (c^2\Lambda^2/mk)\}}\right]}$$

When Λ is low, x_0/X tends to unity; when Λ is high, x_0/X tends to zero. Between these two extremes, the response is dependent on the amount of damping in the system. If c = 0, amplitude could be infinite with a phase lag of $\pi/2$ at the resonant frequency. At the resonant frequency the amplitude of vibration can be very large. For a lightly damped system, the maximum response would occur near the natural frequency of the system. The general response of this system is shown in Fig 11.3.

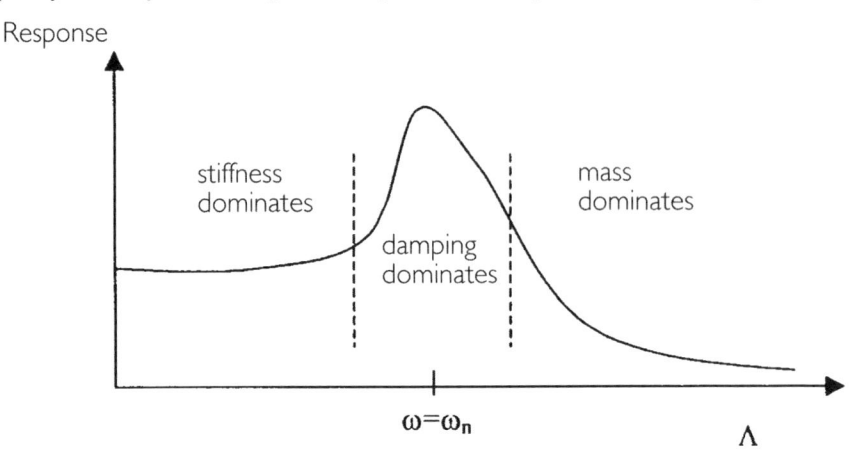

Fig 11.3: Frequency response curve

While it would be undesirable for a ship, or its sub-systems, to undergo resonant vibration, sometimes it may not be easy to avoid such an occurrence and its effects need to be estimated. This can be achieved by examining the system characteristics, such as damping and natural frequency. In practice, many systems can have more than one natural frequency, see Fig 11.4, where a simply-supported beam is undergoing flexural vibration in the vertical plane. With each different number of nodes, there is a corresponding mode of vibration, each with its own resonant frequency.

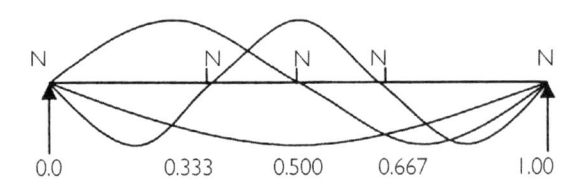

Fig 11.4: Different mode shapes for a simply-supported beam

For irregular excitation, for instance by random wave forces, the forcing function can be represented by a Fourier series:

$$F = \sum_{n=1}^{n=\infty} F_n \sin(n\omega t + \delta_n)$$

The total force is expressed as the sum of its regular components. By the principle of linear superposition, the total vibratory response is then the sum of system responses to each of these components. In practice, only a limited number of terms would be considered. Readers are referred to Den Hartog[1] for a detailed analysis of mechanical vibrations.

11.2 Ship vibration

A ship is an elastic structure. When it is subject to oscillatory forces, which may originate from an internal or external source of excitation, vibration can result. The main engine, auxiliary machinery and shafting excitation are typical internal causes while the propeller and sea waves are the major external causes of vibration.

Internal causes generally arise from mechanical devices interacting directly with the ship structure. For example, a diesel engine with its reciprocating masses could generate large unbalanced forces and moments which, when transmitted to the hull, can excite vibration. The magnitude of excitation depends on the number of cylinders, the firing order and other characteristics. More importantly, their excitation frequencies can be of the same order as the hull vibrations. With greater use of slow-speed engines having fewer cylinders, this type of prime mover is becoming the primary excitation source of global vibrations. To reduce their adverse effects, flexible mountings or compensators are often used. In contrast, rotating machinery such as turbines, do not produce significant vibration excitation.

Auxiliary machinery, such as engines and pumps, are usually run at much higher speeds and their unbalanced forces would occur at a higher frequency. Therefore, low frequency main hull vibrations would not be excited. However, they may excite local vibrations, for instance, in nearby bulkheads. Shafting vibrations, another local phenomenon, could be in the form of axial, whirling or torsional vibrations, which are mainly related to propeller forces. Often, shaft vibrations are attributable to shaft misalignment and out-of-balance rotation.

External causes generally arise from variations in fluid pressure around the hull, eg fluid-structure interaction. For example, a rotating propeller creates an oscillatory pressure field that acts on the hull surface in the propeller vicinity. This pressure in the vertical direction is generally considered the most significant, and hull panels above the propeller can easily suffer fatigue damage. The frequency of these hydrodynamic forces is dependent on the number of blades and the rotation speed. For instance, for a four-bladed propeller running at 120rev/min, the fundamental frequency is 8Hz. The intensity of the pressure fluctuations is affected by propeller characteristics, wake and cavitation effects. A cavitating propeller operating in a non-uniform wake field creates the worst impacts. In this case, the pressure forces can be many orders of magnitudes greater than other propeller problems. It should be noted that a damaged propeller can also cause excessive vibrations, since it will be out of balance. Methods used to reduce these adverse effects include, setting a minimum propeller tip clearance, using a skewed propeller, modifying the loading distribution of the propeller during design and improving the water flow at the stern of the ship.

When a ship travels through waves, it is subject to varying hull pressures. When considering sea-keeping, the resulting rigid body motions are of interest. However, the waves could also excite main hull vibrations when the wave encounter frequency coincides with the hull natural frequency. Under fairly steady and continuous excitation, the resulting vibration is known as **springing**. This type of vibration mainly affects ships that are long and with low hull natural frequencies. When the ship undergoes slamming, either through bottom or bow flare impact, the resulting wave-excited hull vibration is known as **whipping**. These vibrations can cause additional bending stress that could be significant for fatigue consideration. A 1992 study by Orbeck[2] points to the importance of these effects on modern merchant ship design. Generally, vertical vibrations are most important. However, for some ships, such as container vessels, these wave-induced vibrations could cause appreciable horizontal and torsional vibrations. Reducing the ship's speed, or altering course, can reduce these forms of vibration.

In the preceding sections, major excitation sources or disturbing forces acting on a ship were introduced. Two main types of vibration response were apparent: global and local. Global vibrations refer to vibration of the main hull or superstructure, and local vibrations refer to vibration of some small portion of the structure, such as decks, bulkheads or web-frames. Local vibrations are difficult to investigate during the design stage, since there are many possibilities for their occurrence. In practice, most of the local vibration-related problems can be remedied fairly easily. Nevertheless, the designer should be aware of the vibration characteristics of typical ship type structures. When considering global vibration, superstructure vibrations for example could cause great discomfort to ship's personnel and damage to equipment. High excitation levels can be experienced, even at frequencies away from the resonant condition.

Main hull vibration is concerned with the structure as a whole. Once designed and built, it would be very difficult and equally expensive to take remedial action against vibration problems. The hull may bend as a beam in the vertical and horizontal planes and vibrations under these modes are known as **flexural** vibrations. When the hull twists about its longitudinal axis, subject to angular displacement, this is known as **torsional** vibration. The vertical flexing is generally considered more important. The hull's natural frequencies are a function of its shape, mass and stiffness distribution. The flexing of a uniform beam will be considered as an introduction to this topic.

11.2.1 Flexural vibrations of a beam

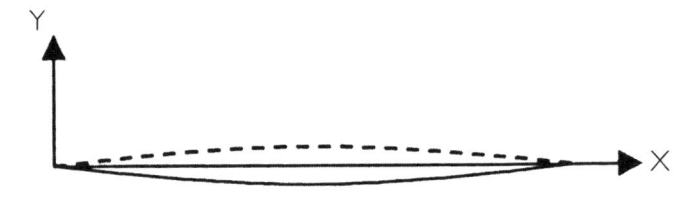

Fig 11.5: A beam undergoing lateral vibration

Consider a beam of uniform cross-section executing lateral vibration in the vertical plane, see Fig 11.5. The deflections are small, and the effects of shear deflections and rotation will be neglected. Vibrations are considered to be simple harmonic motions such that the vertical displacement at any instant at a particular point x along the length of the beam can be described by:

$$\frac{d^2y}{dt^2} = -\omega^2 y$$

where ω (rad/sec) is the frequency of vibration, $2\pi/T$, where T = period.

If M is the mass per unit length, for a length of δx, the force is $M(d^2y/dt^2)\delta x$ and the force per unit length is $M(d^2y/dt^2)$. From beam theory, the equation for the lateral vibration of the beam becomes:

$$IE\frac{d^4y}{dx^4} = -M\frac{d^2y}{dt^2}$$

where I is the second moment of area and E the modulus of elasticity of the beam. Assuming the beam oscillates about the mean position, then

$$\frac{d^4y}{dx^4} = \left(\frac{M}{EI}\right)\omega^2 y$$

Putting $k^4 = (M/EI)\omega^2$, the equation becomes:

$$\frac{d^4y}{dx^4} - k^4 y = 0$$

Hence, once the k values are determined, the natural frequencies of the beam can also be determined. It can be shown that the general solution for this differential equation is of the form:

$$y = A_1\sin(kx) + A_2\cos(kx) + A_3\sinh(kx) + A_4\cosh(kx)$$

Applying boundary conditions for a free-free beam (where the ends are not clamped), for the shear force $d^3y/dx^3 = 0$ at $x = 0,L$; and the bending moment $d^4y/dx^4 = 0$ at $x = 0,L$. The equation reduces to:

$$\cos(kL)\cosh(kL) = 1$$

For the first three modes of vibrations, $k_1L = 4.73$, $k_2L = 7.85$, $k_3L = 10.99$, hence the fundamental frequency is given by $\omega = (4.73)^2\sqrt{(EI/ML^4)}$. Their mode shapes are shown in Fig 11.6, where the fundamental mode of vibration is the two-node vertical. The same analysis can be applied to horizontal vibrations.

For a simply supported beam, the boundary conditions are: deflection = 0, at x=0,L; and the bending moment $d^2y/dx^2 = 0$ at $x = 0,L$. The governing equation becomes $\sin(kL) = 0$ and the fundamental frequency is given by $(\pi)^2\sqrt{(EI/ML^4)}$. Mode shapes are shown in Fig 11.4.

Comparisons of the free-free beam characteristics and measured ship profiles of two-node vertical are given in Table 11.1. It will be seen that real ship vibration profiles are fairly similar to the ideal free-free beam, even though the mass and stiffness distribution varies along the ship. Therefore, this idealised beam can be used as a basis for further analysis of ship vibration problems.

Table 11.1: Comparisons of node locations for two-node vertical measurements

Free-free beam	0.224L(stern)	0.776L(bow)
300 000dwt tanker	0.300L	0.716L
1997TEU containership	0.239L	0.685L

11.2.2 Ship flexural vibrations

The ship, considered as a beam, can flex in the vertical or horizontal plane. The mode of vibration is described by the number of nodes along its length, and the fundamental mode being the two-node mode as shown in Fig 11.6 (a). Each mode has its natural frequency of free vibrations which increase with the number of nodes. In Table 11.2, typical ship vibration frequencies are given. The data, although limited, indicates the following general trend:

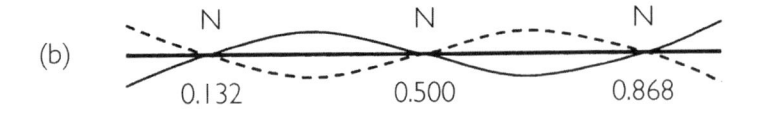

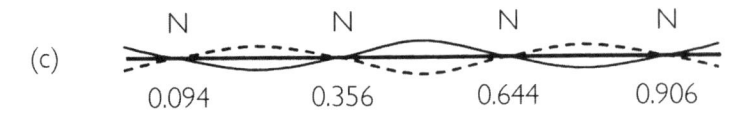

Fig 11.6: Lateral vibration (a) two-node, (b) three-node, (c) four-node

1) Shorter ships would have a higher natural frequency.
2) Ships under full load condition have a lower natural frequency when compared to light ship condition values.

These findings are consistent with the simple frequency formula for the free-free beam. Also, the horizontal frequencies are, in general, about 50% higher than the respective vertical values, reflecting higher horizontal stiffness.

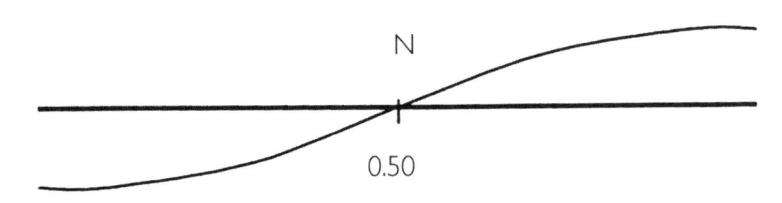

Fig 11.7: One-node torsional vibration, angular displacement

For angular displacement, ie twisting of the hull, torsional vibration will occur. In this case, the fundamental mode is the one-node vibration, see Fig 11.7. Higher modes are also possible. So far, different vibration modes have been introduced as independent of one another. In practice, one vibration mode may generate vibration in another mode. When this happens, the vibrations are said to be coupled. Often, vibration in the horizontal plane would exhibit **coupled** motion (horizontal-torsional).

Table 11.2: Typical ship vibration frequencies

Ship type	Length L_{PP} (m)	Loading	Frequency of Vibration (Hz)							
			2NV	3NV	4NV	5NV	2NH	3NH	4NH	T
Tanker	155	Full	1.13	2.37			1.86			
		Light	1.25	2.72			2.35			
Containership	197	Light	1.03	2.05	2.93	3.76	1.37	2.84	3.86	0.88
Bulk carrier	200	Light	0.99	1.92	2.78	3.55	1.73	3.60	4.80	4.00
Passenger ship	171	Full	1.58	2.72	3.90	5.00	1.97	4.27		
Tug	60	Full	4.18							
Ice-breaker	76	-	4.33	9.00	12.0		7.00			
Trawler	50	Full	4.83	9.57	13.9		6.98	13.9		

11.2.3 Ship vibration formulae

It has been shown that a formula for ship vibration could take on a similar form to that of a free-free beam. Obviously, the ship will have a varying distribution of mass as well as second moment of area (stiffness) along its length. However, if these properties amidships could be taken as characteristic of a beam-like ship, then for the ship:

$$N = \text{constant} \times \sqrt{(I_{\varnothing}/\Delta L^3)}$$

where the mass term is replaced by the displacement of the vessel, length of beam by length between perpendiculars of the ship, and the second moment of area by the midships section value, $I_{\varnothing}$. The modulus of elasticity is assumed unchanged, as for ships, it is mostly that of steel. For non-steel vessels, the constant would take a different value.

This type of formula is known as the Schlick formula, after Otto Schlick, who was one of the earliest proponents of ship vibrations. To obtain values for the constant, tests were normally performed on the ship by exciting the hull over a range of frequencies. As the natural frequencies vary with displacement, data would be needed for different loading conditions. Stiffness distribution also affects this constant, which needs to be derived for different ship types. Finally, in addition to these two factors, which have a direct bearing on frequency, the following need further examination:

1) Added mass-influence of entrained water.
2) Departure from ordinary bending theory.

11.2.4 Added mass

When a beam vibrates in a fluid, the total kinetic energy of the system is the sum of the kinetic energy of the beam and of the surrounding fluid. For a beam vibrating in air, the kinetic energy of the air could be ignored because of its low density. However, for a ship's hull, which is partially immersed in water, the kinetic energy of the surrounding water cannot be ignored as its density is much higher. The amount of fluid being set into motion will have an effect on the frequency of the vibrating system.

It can be shown that when a circular cylinder vibrates in water, the total kinetic energy of the system is increased by an amount proportional to the volume of fluid displaced by the cylinder, ie for a cylinder length L of radius r, it is $\pi r^2 L$ at a particular velocity. The equivalent mass is therefore $\rho\pi r^2 L$. The net effect is an increase in mass of the system. This is often referred to as added mass or added virtual mass. It shows that, for a cylinder vibrating in water, the **added mass** effect is equivalent to doubling the displacement of the system. With an increase in mass, the natural frequency of the system would be lower than for the same system in air.

For a typical ship section, it would be reasonable to expect the direction of motion and hull shape to have an influence on added mass. The hull shape influence can be determined based on the breadth (B) and draught (T) of the section, for instance (B/T). Based on the cylinder example, for a half-immersed section, the added mass of a ship would be:

$\int \rho A(B/2T)\, dx$

where A is the immersed section area.

The total virtual mass Δ_1, in terms of ship's breadth (B) and draught (T) for vertical modes can be expressed as:

$\Delta_1 = \Delta(1 + k\, B/T)$,

where Δ is the displacement of the ship and k is a factor to be determined.

Two commonly-accepted formulae for vertical and horizontal vibration of ships are:

$\Delta_1 = \Delta(1.2 + B/3T)$ vertical

$\Delta_1 = \Delta + 0.574\, LT^2$ horizontal

More sophisticated mathematical models for calculating added mass are available, for example, ship-like sections based on a conformal mapping technique by Lewis[3]. For these sections, which are based on two-dimensional analysis, a correction factor for the end effects, ie three-dimensional effects, is also introduced. For further information, readers are referred to Lewis.[3]

11.2.5 Bending theory

Under the assumption of pure bending, sections that are plane and perpendicular to the neutral axis remain so after bending. The beam under this condition can be considered as a series of rectangular elements. In practice, shear stress exists and these rectangular elements would tend to go into a diamond shape, forming shear angles. The additional deflection is referred to as shear deflection. Also, the total stress is no longer a simple function of distance from the neutral axis. By considering a simple-supported beam with a concentrated mass at mid span, it can be shown that the vibration frequency of the beam is reduced due the presence of shear deflection.

For ship vibration, a proposed shear correction factor is:

$r_s = C \times 100 \times D^2/L^2$

where C is a constant dependent on the breadth to depth ratio, and D is the depth. Research work suggests that the correction factor is:

$$r_s = 2.5\left(\frac{D^2}{L^2}\right)\left\{\frac{3\left(\frac{B}{D}\right)^3 + 9\left(\frac{B}{D}\right)^3 + 12\left(\frac{B}{D}\right) + 1.2}{3\left(\frac{B}{D}\right) + 1}\right\},$$

and the frequency reduction is given by:

$$\frac{1}{(1+r_s)^{0.5}}$$

In general, the reduction is about 10% for two-node vertical and can be much greater for higher modes.

As the depth of the ship increases, kinetic energy of rotation would also increase. This arises as sections further away from the neutral axis would experience greater rotation of mass about a transverse axis. This is shown in Fig 11.8. The net effect is again a reduction in frequency. For two-node vertical, this reduction is about 2–3% but would increase to 8–10% for four-node vertical.

Fig 11.8: Rotation effect

For ships with extensive superstructures, the application of bending theory would need correction too. Here, the stress distribution no longer varies linearly with depth. To account for this, an effective value of second moment of area, I, is often estimated based on semi-empirical formulae.

11.3 Approximate formulae

In the early design stages, there is a need to establish the natural frequencies of a ship. However, detailed information on mass and stiffness (represented by the second moment of area) distribution may not be available. In practice, it would be necessary to rely on simple empirical formulae for initial estimates. Although the Schlick type formula has its limitations, it has provided a sound basis for practical use.

There have been many modifications to the Schlick formula, for instance, Burrill[4] who introduced factors for added virtual mass and shear deflection. As his analysis was based on ship designs of the 1930s, the derived data is only of historical interest today. More recent variations of the Schlick type formula are introduced in the following sections.

11.3.1 Kumai's approach

Based on Kumai's formula, the following expression for two-node vertical vibration is given in Johannessen and Skaar[5]:

$$N_{2v} = 1.62 \times 10^6 \sqrt{\left(\frac{I}{\Delta_1 L^3}\right)} \text{ Hz}$$

where I is the second moment of area (m⁴);
$\Delta_1 = \Delta\{(B/3T)+1.2\}$, displacement including virtual added mass (kg);
L = length between perpendiculars (m);
B = beam amidships (m);
T = mean draught (m).

Based on comparisons with finite element methods, accuracy in the range of ±10% could be expected.

Another general formula, derived from observed results by other researchers, is frequently used for two-node vertical vibrations:

$$N_{2v} = \frac{\left\{238\,660\sqrt{\left(\frac{I}{\Delta_1 L^3}\right)} + 29\right\}}{60} \text{ Hz}$$

where I is in m⁴ and the displacement term is in MN and includes the added mass effect.

Example

A bulk carrier, 200m in length and 27m in beam, has a draught of 6.2m. Given that the second moment of area of the midship section is 142.8m⁴ and a displacement of 26 000 tonnes, estimate the two-node vertical vibration frequency.

$$\Delta_1 = \Delta\left\{\left(\frac{B}{3T}\right)+1.2\right\} = 26\,000 \times \left\{\left(\frac{27}{3\times 6.2}\right)+1.2\right\} = 68\,940 \text{ tonnes}$$

Using

$$N_{2v} = 1.62 \times 10^6 \sqrt{\left(\frac{I}{\Delta_1 L^3}\right)} \text{ Hz}$$

$$N_{2v} = 1.62 \times 10^6 \sqrt{\left(\frac{142.8}{(68.94 \times 10^9 \times 200^3)}\right)} \text{ Hz}$$

$$N_{2v} = 0.824 \text{ Hz}$$

Using

$$N_{2v} = \frac{\left\{238\,660\sqrt{\left(\frac{I}{\Delta_1 L^3}\right)} + 29\right\}}{60} \text{ Hz}$$

$$N_{2v} = \frac{\left\{238\,660\sqrt{\left(\frac{142.8}{676.3 \times 200^3}\right)} + 29\right\}}{60} \text{ Hz}$$

$$N_{2v} = 1.129 \text{ Hz}$$

From actual measurement, the two-node vertical frequency was 0.99Hz. Therefore the difference was from -17% to +14% respectively, which would be a fairly reasonable first attempt, when no additional information was available.

Example

A ship of length 134m, breadth 19.7m and draught 7.58m has a displacement of 15 600 tonnes. The second moment of area, I, is 36.4m⁴. Determine the natural frequency of vibration allowing for added mass effects.

$$\Delta_1 = \Delta\left\{\left(\frac{B}{3T}\right) + 1.2\right\} = 15\,600 \times \left\{\left(\frac{19.7}{3 \times 7.58}\right) + 1.2\right\} = 32\,300 \text{ tonnes}$$

Using

$$N_{2V} = 1.62 \times 10^6 \sqrt{\left(\frac{I}{\Delta_1 L^3}\right)} \text{ Hz}$$

$$N_{2V} = 1.62 \times 10^6 \sqrt{\left(\frac{36.4}{32.3 \times 10^6 \times 134^3}\right)} \text{ Hz}$$

$$N_{2V} = 1.11 \text{Hz}.$$

Using

$$N_{2V} = \frac{238\,660\sqrt{\left(\frac{I}{\Delta_1 L^3}\right)} + 29}{60} \text{ Hz}$$

$$N_{2V} = \frac{238\,660\sqrt{\frac{36.4}{\left(316.9 \times 134^3\right)}} + 29}{60} \text{ Hz}$$

$$N_{2V} = 1.35 \text{Hz}.$$

Example

A single-deck ship has a length of 122m, breadth 16.75m, depth 10.95m and, with a block coefficient of 0.7, floats at a uniform draught of 7.32m. The midship section of the ship can be assumed rectangular and the vessel has a double bottom 1.07m in depth with a centre girder. All the material can be assumed 12.5mm thick. Use the two approximate formulae to estimate the natural frequency of vertical vibration.

Item	Sectional area A $(m^2) \times 10^{-4}$	Lever (m)	Moment $\times 10^{-4}$ $(m^2)(m)$	Lever (m)	I $(m^4) \times 10^{-4}$	$Ah^2/12$ $(m^4) \times 10^{-4}$
Deck	2100	10.95	23 000	10.95	252 000	-
Tank Top	2100	1.07	2240	1.07	2400	-
Bottom	2100	-	-	-	-	-
Sides	2740	5.48	15 050	5.48	82 500	27 400
Centre girder	134	0.53	71	0.53	38	13
	9174		40 341		336 938	27 413
					27 413	
					364 351	

Neutral axis above the base = $4.0360 \div 0.9174 = 4.4$ m

I about the base = 36.44 m^4
I about NA = $36.44 - (0.9174 \times 4.4^2)$
       = 18.74 m^4

$\Delta = 122 \times 16.75 \times 7.32 \times 0.7 \times 1.025$ t
  = 10 700 tonnes

$$\Delta_1 = \Delta \left\{ \left(\frac{B}{3T} \right) + 1.2 \right\}$$

$$= 10\,700 \times \left\{ \left(\frac{16.75}{3 \times 7.32} \right) + 1.2 \right\} = 21\,000 \text{ tonnes}$$

Using

$$N_{2V} = 1.62 \times 10^6 \sqrt{\left(\frac{I}{\Delta_1 L^3} \right)} \text{ Hz}$$

$$N_{2V} = 1.62 \times 10^6 \sqrt{\left(\frac{18.67}{21 \times 10^6 \times 122^3} \right)} \text{ Hz}$$

$N_{2V} = 1.134$ Hz

Using

$$N_{2V} = \frac{\left\{ 238\,660 \sqrt{\left(\frac{I}{\Delta_1 L^3} \right)} + 29 \right\}}{60} \text{ Hz}$$

$$N_{2V} = \frac{238\,660\sqrt{\left(\dfrac{18.67}{206 \times 122^3}\right)} + 29}{60} \quad \text{Hz}$$

$$N_{2V} = 1.37 \text{Hz}.$$

Using Shlick's original formula, the estimated frequency is 1.47Hz.

11.3.2 Todd's approach

In the early design stages, it would be more convenient to make use of basic parameters, such as breadth and depth, for preliminary calculations before data on second moment of area is available. In this context, Todd[6] proposed that the relationship $I = BD^3/12$ can be taken as $I = C(BD^3)$, where C is a coefficient which varies with ship type and geometry. This idea is reasonable, particularly for ships with box type sections. The Schlick formula would be modified to:

$$N = \beta\sqrt{\frac{BD^3}{\Delta L^3}}$$

and to take into account of added mass effect, $\Delta_1 = \Delta\left\{\left(\dfrac{B}{3T}\right) + 1.2\right\}$, the expression becomes:

$$N = \beta\sqrt{\left(\frac{BD^3}{\Delta_1 L^3}\right)}$$

Typical β values for tankers are: large tanker (fully loaded) – 11 000; small tankers (fully loaded) – 8150. The displacement in this formula would be in MN, linear dimensions in metres and N in cycles per minute. This formula has been used as a basis of analysis for different ship types and found to give similar results to those based on second moment of area.

When a superstructure extends in excess of 40% of ship length, it would have an influence on vibration frequencies. The concept of effective depth was thus introduced to take into account the influence of superstructure on sectional properties.

With reference to Fig 11.9, the following expression is applicable:

$$\text{effective depth, } D_E = \sqrt[3]{\{D^3(1 - x_1) + D_1^3(x_1 - x_2) + D_2^3 x_2\}}$$

where D is the depth of the uppermost continuous deck. Other correction factors, which take into account of stress distribution of superstructures, have also been proposed by Johnson and Ayling[7].

11.3.3 Comparative approach

A reliable method for estimating the natural frequencies of a new design is on a comparative basis. This method uses data from geometrically-similar ships under the same loading conditions as the starting point. For instance, if I_1 is the second moment amidships, Δ_1 is the displacement including added mass and L_1 the ship length between perpendiculars of the new design and I_2, Δ_2 and L_2 are particulars of the similar ship, the relationship for this method is:

Fig.11.9: Effective depth

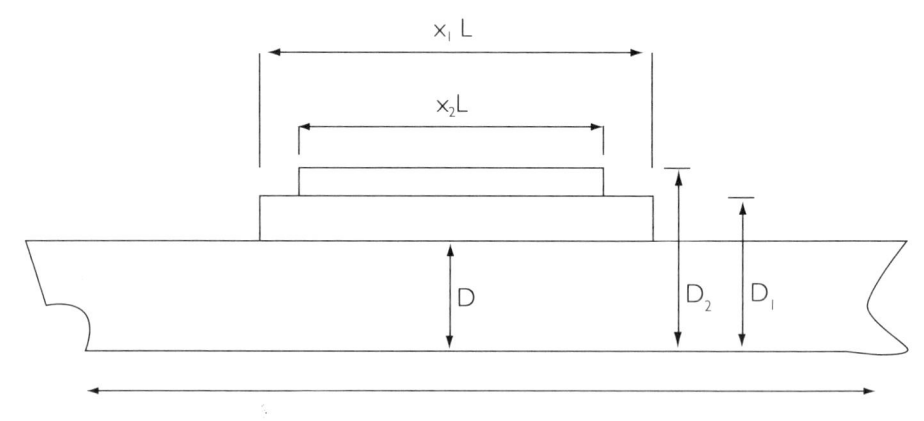

L = Length between perpendiculars

$$\frac{N_1}{\sqrt{\left(\frac{I_1}{\Delta_1 L_1^3}\right)}} = \frac{N_2}{\sqrt{\left(\frac{I_2}{\Delta_2 L_2^3}\right)}}$$

Hence,

$$N_2 = N_1 \sqrt{\left\{\left(\frac{I_2}{I_1}\right)\left(\frac{\Delta_1}{\Delta_2}\right)\left(\frac{L_1^3}{L_2^3}\right)\right\}}$$

This relationship can be used to estimate both vertical and horizontal hull resonant frequencies. In practice, these estimates can have an accuracy of over 95% to measured values of the lowest modes of vibration. Following Todd's proposal, the second moment of area, I, term can be replaced by BD^3.

Similarly, for torsional vibration, the expression is:

$$\frac{N_1}{\sqrt{\left[\frac{J_1}{\Delta_1}\left(B_1^2 + D_1^2\right)L_1\right]}} = \frac{N_2}{\sqrt{\left[\frac{J_2}{\Delta_2}\left(B_2^2 + D_2^2\right)L_2\right]}}$$

where J_1 is the torsional constant, B_1 is the moulded breadth, D_1 is the moulded depth and Δ_1 is the displacement but with added mass of the new design and J_2, B_2, D_2, Δ_2 are the corresponding particulars of the similar ship. The accuracy of the estimate is not as high as the corresponding vertical mode estimate but it is considered sufficient for locating frequencies that are likely to pose problems.

Example

The hull of a vessel 150m in length, 20m in breadth and 12m depth, floats at a draught of 9m with a displacement of 150MN and a tonne per centimetre immersion,

TPC, of 18. The two-node vertical frequency of vibration, including the effect of added virtual mass, is 1.39Hz.

A superstructure 40m long, 3m high and weighing 400tonnes is then fitted. Determine the new frequency of vibration, including the effect of added virtual mass.

Total virtual displacement, $\Delta_1 = \Delta\left\{\left(\dfrac{B}{3T}\right) + 1.2\right\}$

$= 150\left(\dfrac{20}{3 \times 12} + 1.2\right)$

$= 263.3\text{MN}$

After adding superstructure:

New draught = $12 + \left(\dfrac{400}{18 \times 100}\right) = 12.2\text{m}$

New displacement = $150 + (400 \times 9.81 \times 10^{-3}) = 153.9\text{MN}$

New virtual displacement = $\Delta_2 = 153.9 \times \left\{\left(\dfrac{20}{3 \times 12.22}\right) + 1.2\right\}$

$= 268.6\text{MN}$

Effective depth, $D_E = \sqrt[3]{\left\{D^3(1-x_1) + D_1^3(x_1-x_2) + D_2^3 x_2\right\}}$

$= \sqrt[3]{\left\{12^3\left(1 - \dfrac{40}{150}\right) + 15^3\left(\dfrac{14}{150}\right)\right\}}$

$= 12.94\text{m}$

New frequency of vibration, $N_2 = 1.54\text{Hz}$

11.3.4 Higher orders

Higher order vibration frequencies are also needed for preliminary design purposes, for example to locate frequencies that may fall close to main engine operating frequency. The frequency of interest in this case, would be at least four to five nodes. These frequencies could be estimated based on the two-node value by the following formula proposed in Johannessen and Skaar[5].

$N_{NV} \cong N_{2V}(n-1)^\mu$

where n is the number of nodes and should not exceed five or six. The index μ = 0.845 for general cargo ships, 1.0 for bulk carriers and 1.02 for tankers. Comparisons with earlier data by Johnson and Ayling[7] are given in Table 11.3. In higher modes, shear effects become increasingly important. The good comparisons may be fortuitous, but they do demonstrate the value of the method. In general, the approximation should be used with care and compared with data from similar ships.

Table 11.3 Higher order frequency (Hz) comparisons

Tanker	2NV	3NV	4NV	5NV	6NV	7NV
Data (best line)	1	2.10	3.15	4.20	5.20	6.25
Formula	1	2.03	3.07	4.11	5.16	6.22

11.4 Direct calculation methods

Empirical formulae are useful to derive estimates for preliminary design purposes, but their accuracy is limited by the amount of data available. For new ship types, or less conventional ships, this approach would not be adequate. Nowadays, with high power computers, finite element methods are widely used by designers. Based on the structural definition of a ship, vibration frequencies of different modes and orders can be analysed. This would involve a fairly full description of the structural arrangement of a ship, which would be at an advanced stage of design. For the present purpose, it is instructive to introduce two direct calculation methods to gain an insight into the subject. These are the deflection, or full integral method, and the energy method. They can produce estimates for two-node frequency with reasonable accuracy for practical application.

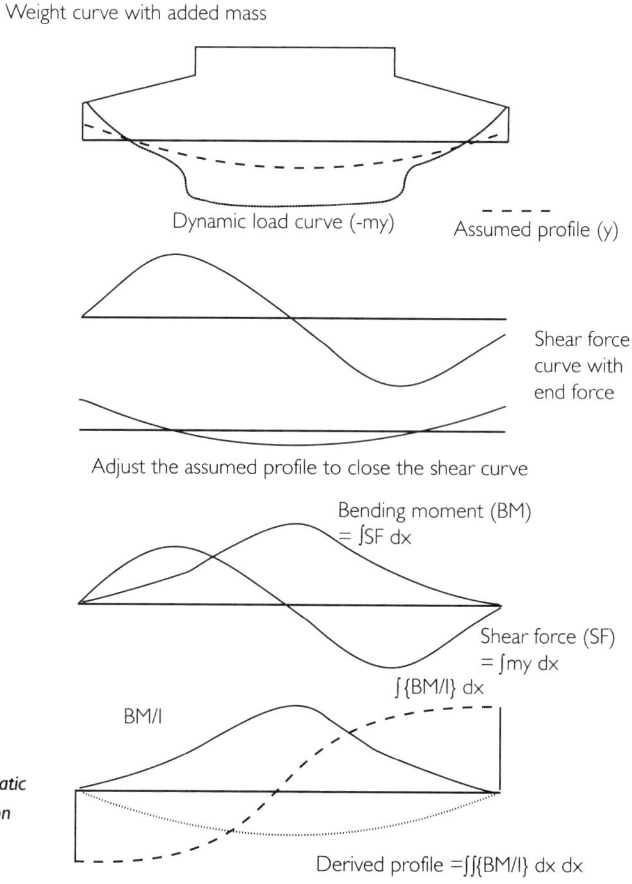

Fig 11.10: Schematic layout of deflection method

11.4.1 Deflection method

In this method, the ship is assumed to undergo simple harmonic motion, where the deflection profile is given by $y = f(x)\sin\omega t$. The acceleration value of this function is given by differentiating y twice with respect to t and the dynamic loading becomes $-m\omega^2 y$, where m is the mass at the particular point concerned. An initial deflection profile is based on that of a free-free beam.

Further integration of the dynamic loading leads to integral expressions for shear force ($-\omega^2 \int m\ y^2\ dx$) and bending moment ($-\omega^2 \iint m\ y^2\ dx\ dx$), the constants of integration being zero, as shear force and bending moment are zero at the origin. Based on the bending moment and deflection relations, corresponding integrals for slope and deflection can be derived, and the actual deflection y_a is given by:

$$y_a = \iint (1/EI)\ \omega^2\ \{\iint m\ y^2\ dx\ dx\ \}\ dx\ dx$$

Thus, by integrating the dynamic load four times, the actual deflection can be derived. Based on a comparison of the assumed and actual deflection profiles, eg at amidships, the frequency of vibration can be deduced.

In practice, there are complications to this approach. The mass distribution along the ship length, including added mass, is not uniformly distributed. As a result, the shear force curve would not close at the ends. This difficulty is overcome by shifting the base-line until the shear force at the end points are close to zero, then further integration would produce a deflection curve. If the derived profile is significantly different from the assumed profile, then the derived curve will be used as the new input assumed profile, starting an iteration process. The calculation process is illustrated schematically in Fig 11.10.

11.4.2 Energy method

Based on the principle of conservation of energy, assuming damping is negligible, it can be stated that the total energy of a vibrating system is constant. Hence, the sum of kinetic energy and potential energy of a vibrating system can be considered constant. For a vibrating beam, the moving masses undergoing linear motion generate kinetic energy and the potential energy is the strain energy of bending, when the beam deflects from its rest position.

From beam theory, it can be shown that:

$$\text{strain energy of bending} = \frac{1}{2}\int EI\left(\frac{d^2 y}{dx^2}\right)^2 dx$$

where the integration is carried out over the entire beam. The potential energy would be at a maximum when the beam is deflected to its extreme position, when the velocity becomes zero and it is about to return to its rest position.

The kinetic energy for the beam, with mass per unit length, m, undergoing simple harmonic motion is:

$$\text{kinetic energy} = \frac{1}{2}\int m\left(\frac{d^2 y}{dt^2}\right)^2 dx = \frac{1}{2}\omega^2 \int my^2 dx$$

where y is an assumed deflection profile. When the beam passes through its rest position, when no bending occurs, its velocity will be at a maximum producing a maximum in kinetic energy.

The total energies of the system being constant, equating the two energies, the frequency of vibration, ω, of the beam is given by:

$$\omega^2 = \frac{\int EI \left(\frac{d^2y}{dx^2}\right)^2 dx}{\int m y^2 dx}$$

Applying the above to a simply-supported beam and assuming that

$y = y_0 \sin\left(\frac{\pi x}{L}\right) \sin \omega t$, it can be shown that the fundamental frequency, ω,

is $\pi^2 \sqrt{\frac{EI}{mL^4}}$. This is the same as the exact solution shown in the previous section.

Hence, if the assumed deflection profile is close to the actual deflection and system damping can be ignored, this method can yield reasonable estimates. In application to the case of a ship, again, an initial deflection profile could use the 'free-free' beam. From ship measurements, damping is about 1–3.5% of critical damping and could be considered negligible for these calculations.

For analysis of local structures, such as masts, this method has been used to estimate the first natural frequency of vibration by modelling the mast as a cantilevered beam. A similar expression to the above can be derived. For a cantilever with a concentrated mass at its end, an approximation to the equation is:

$$\text{frequency,} \quad f = 0.276 \sqrt{\frac{EIg}{(M + 0.23m)L^3}} \text{ Hz}$$

where M is the end mass, and m the mass of the beam.

11.4.3 Finite element method

From the two direct methods above, it might be expected that frequencies of higher modes could be obtained if appropriate deflection profiles are at hand. However, in practice, the derived profiles tend to converge to the two-node profile and it is not possible to derive even the three-node mode. Although methods to overcome this divergence problem have been proposed by earlier researchers, today's analytical methods are mainly based on finite element analysis. A brief outline of this approach based on the beam model for ship vibration is given in the following section.

The beam model is the simplest finite element model which would be used for initial assessment. The ship structure is represented by an array of finite beam elements along the main hull axis. Each element is assigned the mass and stiffness properties derived from the ship's structure, loading and added mass effect. For two-node vertical frequencies, between 20 to 30 elements have shown to be sufficient for the first two modes of vertical vibration. In general, this model can be used with confidence for the prediction of natural frequencies up to about eight-node vertical, depending on ship type. This method can be extended to calculate horizontal and torsional vibrations, but the scope would be limited to three–four node horizontal and one-node torsional only.

Two-dimensional and three-dimensional models can further enhance the accuracy and scope of the analysis, including coupled effects, such as horizontal-torsional vibrations. Realistic results can be obtained for natural frequencies up to and beyond 15-node vertical vibrations.

11.5 Noise

Noise is closely related to ship vibration. Noise can cause a degradation of crew performance, since they are exposed to shipboard noise 24 hours a day. The quality of life onboard can be badly affected and, in extreme cases, hearing loss can result from long-term exposure to excessive noise. On a passenger ship, noise is highly undesirable as it causes discomfort and nuisance.

Similarly to ship vibration, there are various causes of noise and they may share a common origin:
- Propeller-structural and airborne noise, emanating from the aft part of a vessel due to alternating propeller forces (sometimes arising from cavitation).
- Machinery-structural and airborne noise generated by vibration of a machine's foundation, or winches squeaking at a high pitch.
- Fluid flow-structural and airborne noise, caused by the flow of air in ducting systems, steam and water in the piping systems, fuel oil transfer or cargo pumping.
- Electrical component-airborne noise, from transformers and generators.

Structural-borne noise refers to structural vibration in the frequency range 16Hz–20kHz, which radiates through a fluid medium, such as air and water. Its origin is mainly mechanical in nature. Its effects depend upon the source strength, transmission properties and radiation properties of the structure at the receiving end. Having low internal damping, vibrations within ships will travel a long way. As soon as large surfaces – such as hull plating – which act as amplifiers, are connected to the vibrating structure, a high noise level results.

11.5.1 Sound measurement

The magnitude of sound is often measured in terms of its pressure, which in SI unit is N/m² or Pa. The threshold of human hearing has been found to be about 20×10^{-6} Pa. For sound level measurements, this value has been used as a reference level. A convenient scale for acoustic measurements can be defined as:

$$\text{sound pressure level} \quad L_p = 10 \log_{10}\left(\frac{p}{p_0}\right)^2 = 20 \log_{10}\left(\frac{p}{p_0}\right)$$

where p_0 is the reference level and p is the sound being measured. The sound pressure levels are expressed in *decibels*, dB. For instance, at 1000Hz, normal speech is about 60dB and wood-sawing is about 100dB, relative to 20×10^{-6} Pa. It should be noted that a zero decibel level does not mean that there is no noise; simply that the measured level is the same as the reference.

In terms of acoustic power, the sound power level is:

$$\text{sound power level,} \quad L_w = 10 \log_{10}\left(\frac{W}{W_0}\right)$$

where W is the power emitted and W_0 is the reference power, usually taken as 10^{-12} watts. For instance, a chipping hammer would have a sound power of about one watt and 120dB, based on the above power reference level. Directly related to sound power is a fundamental acoustic quantity known as **sound intensity** (W/m²), which is a vector quantity that describes the amount and direction of net flow of energy at a given position.

In the open air, sound intensity decreases in proportion to the square of the distance from the point source. In decibel terms, doubling the distance from a point source will reduce the dB level by six, ie $20 \log_{10} 2$. For two equal point sources emitting sound, the combined effect is not twice the dB level, but an increase of 3dB. In general, subjective effects of changes in noise levels have been found to be as follows: a change of 3dB in noise level would be just perceptible, for 5dB change, clearly perceptible and for 10dB change it would appear twice as loud. This subjectivity arises as a noise typically contains many components of different frequency, which affect individuals differently. For subjective responses, the A-weighted sound level has shown good correlation, and noise level is expressed as dB(A). For design purposes, typical recommended noise levels are given in Table 11.4.

Table 11.4: Acceptable noise levels for merchant ships

Locations	Levels dB(A)
Sleeping cabins	50-55
Day cabins	55-60
Alleyways	70-75
Machinery space (manned)	90
Workshops	85
Wheelhouse	75

11.5.2 Noise prediction

Acoustical calculations, based on statistical energy analysis (SEA), provide a useful tool for the ship designer in noise estimation and its control. SEA evaluates the dynamic response of structures by taking frequency, spatial and time averages, balancing the energy of all sub-structures. In this approach, the ship structure is divided into a system of elements having certain dimensions and physical properties, similar to the finite element model for vibration analysis. A matrix of coupled linear equations is thus set up. Through assigning the external input power, internal power dissipation and transmission of power from one element to another, the structure-borne noise, due to mechanical or acoustical excitation, is predicted. For air-borne noise, general acoustic theory, taking into account surface areas and absorption properties within an enclosed space, can be used. For further details, readers are referred to Beranek and Ver[8].

11.5.3 Noise control

There are three principal approaches to noise control: isolation, damping and absorption. In isolation, ship space arrangements can be made so that buffer zones around noisy areas, such as the engine room, can bring about acoustic attenuation. For the superstructure and machinery, foundations can be placed on resilient supports as a form of isolation from the main structure. In extreme cases, the noise receiver may be isolated through the use of ear protectors. As with vibrations, structural damping can be altered to change its excitation response to noise. This can be brought about by changing the joints in a panel, mass redistribution, varying material selection, or changing the way local structure is stiffened. Absorption is widely used for air-borne noise, generally by way of insulation, enclosure or barriers, for instance; ventilation system silencers, engine room insulation with absorbent walls, and turbochargers shielded by noise protectors. For noise control to be effective, it must be considered early in the ship design process.

11.6 References

1. Den Hartog JP. *Mechanical Vibration*, McGraw Hill Book Co., 1956.
2. Orbeck F. *Wave-excited hull vibration reduction using hull vibration damper*, Trans. IMarE, 1992..
3. Lewis FM. *Principle of Naval Architecture* (chapter on Hull Vibration of Ships). SNAME, USA.
4. Burrill LC. *Ship vibration: simple methods of estimating critical frequencies*, Trans. North East Coast Institution of Engineers and Shipbuilders, 1934-1935.
5. Johannessen H, and Skaar KT. *Guidelines for prevention of excessive ship vibration*, Trans. SNAME, Vol. 88, 1980.
6. Todd FH, *Ship Vibration*, Edward Arnold, 1961.
7. Johnson AJ, and Ayling PW. *Graphical presentation of hull frequency data and the influence of deck houses on frequency prediction*, Trans North East Coast Institution of Engineers and Shipbuilders, 1956-1957.
8. Beranek L, and Ver I. *Noise and Vibration Control Engineering: Principles and Applications*, John Wiley & Sons, 1992.

12 Ship design

The preceding chapters are an introduction to the different elements of basic ship design. Structural analysis is undertaken to ensure that the ship, with a minimum light-ship weight, is strong enough to withstand all external and internal loads likely to be encountered in its lifetime. Stability calculations check the vessel's ability to float upright, even under extreme operating conditions. Resistance has to be as low as possible to ensure economical propulsion. The synthesis of these different design features enables a ship of particular dimensions and layout to fulfil specific design objectives. In practice, not only will the design need to meet the prevailing technical standards and regulations, but also, it must meet economic objectives of cost-effective operation. This chapter brings together these different aspects, and provides an overview of the ship design process.

12.1 Design objectives

A merchant ship's primary function is to safely convey cargo, which may include passengers, from one port to another, where other transport modes are either uneconomical, or impractical. Although competition from other transport modes, eg air or land (via bridges or tunnels) is increasing, sea transport remains an important means for the international movement of goods and raw materials. Shipowners see a ship as a transport service and an investment, upon which they expect a return. In the maritime transport supply chain, a naval architect's role is, therefore, to provide the ship operator with a safe, efficient and environmentally-friendly ship for the owner's business venture.

Ships operate in a hostile environment in which wind, waves, extreme temperatures and even the water in which they float could be potentially hazardous (corrosive and sometimes as ice). Under these varying operating conditions, the ship must protect its cargo, sustain its speed, reach the next port of call on time, discharge and then load cargo and sail on. The cargo has to be secure and not damaged, whether by excessive motions, flooding or even fire. Should any mishap occur on board, it must be contained immediately. For the months that the ship is at sea, all engineering systems must work reliably and be readily maintained by the small number of shipboard personnel. For the crew, the ship is not only a work place, but also their home, providing for all their needs for weeks at a time. The ship must offer a reasonable level of comfort, and be free from excessive vibration or noise. On a passenger ship, the level of comfort will often exceed that of a five-star hotel. For a successful design, apart from these functional considerations, the ship's aesthetic appeal may be just as important as its structural integrity or hydrodynamic performance.

Over the past two decades, catastrophic accidents at sea, such as the loss of the passenger ferry, *Herald of Free Enterprise*, the grounding of the supertanker, *Exxon Valdez*, and the loss of the bulk carrier, *Kowloon Bridge*, have renewed impetus in the continuing concern for safety of life, the environment and property by the maritime community. New safety rules and regulations have been developed at an accelerated pace by international maritime agencies and organisations. Some of these developments have created new design features in certain ship types, notably the double-hull tanker and the ro-ro ferry. Environmental issues are an increasing concern with respect to air and water pollution. Reducing air pollution will require the use of better quality fuel, which will have implications on engine and fuel system design or it may initiate a general trend towards energy-efficient ships that use a higher grade of fuel. It may even renew interest in wind-assisted ships. The success of high-speed ferries in recent years has brought with it impacts to the environment. High-speed ferry wash can cause damage to sensitive shore-lines and pose a potential danger to those undertaking recreational activities along the coast. A new design goal for such vessels will be to create minimum wash.

In terms of safety, SOLAS is the major international convention which has influenced design as well as operational aspects of ships. From a design perspective, intact and damage stability are two of the key aspects of ship safety. The ship, having satisfied these criteria, will be required to float upright under normal operating conditions and be able to remain afloat in a stable position when damaged. Since the *Herald of Free Enterprise* incident, one of the design requirements of a passenger ship in the damaged condition is that 'the maximum angle of heel after flooding, but before equalisation shall not exceed 15deg'. This has significantly changed the design of passenger ships, and the determination of subdivision. The designer may also be interested to find out what would happen if, in addition, this damaged ship is exposed to environmental forces such as wind or waves.

Structural integrity of the ship is another major safety concern. Notably, the concern for the safety of bulk carriers in recent years has reflected a need to improve structural safety of this ship type. Structural aspects have been an area in which classification societies have played a leading role. They develop rules, based on sound technical principles, which are tested thoroughly against service performance. Compliance with these rules will ensure the provision of adequate overall or 'global' strength, together with adequate local strength of individual components. Design values of still-wave and wave-induced loads are stipulated; scantling and materials requirements are also set out. Whilst the basic strength requirements of ships have not gone through any significant changes in recent years, bulk carrier structural safety has focused the designer's attention on corrosion and fatigue considerations. These aspects are particularly important for older ships and, in general, will also have implications on the design life of the ship. Over recent years classification societies have developed advanced computer programs to assess the structural response to complex dynamic loads.

Increasing concern for environmental protection is reflected in the rising number of proposals for new rules that aim to reduce the pollution risks, particularly by tankers. Regulatory bodies tend to control these risks by the design of specific layouts, such as segregated ballast tanks in tankers. Since the *Exxon Valdez* incident, tankers trading in United States' waters are required to have double hulls. Whilst these initiatives would reduce certain risks, there are also design features that would bring other disadvantages, such as difficulties with hull maintenance. Nonetheless, the major convention for preventing marine pollution, MARPOL 73/78, provides effective measures for controlling pollution.

Over the past two decades, designs for enhancing ship performance in terms of resistance, propulsion, seakeeping and manoeuvrability have been developing at a rapid pace. High-speed ferries now reach speeds in excess of 50 knots, whilst containerships travel at 25 knots with a delivered power under 20MW. These advances are achieved through better hull form design and technological advancements in marine propulsion systems. With greater emphasis on different hull configurations, such as the trimaran and other multi-hull concepts, further increases in speed could be realised. Advances in seakeeping have allowed the operability of ships to extend beyond traditional constraints, through better hull design or the use of stabilising equipment. This has been reflected in designs placing greater emphasis on ride comfort, which is essential for passenger ships. Motion-induced stress for large bulk carriers has been examined using stress-monitoring devices on board ships. In terms of habitability, standards for noise and vibration control are becoming widely accepted practice. With waterjet propulsion systems and thrusters, a ship's manoeuvrability in confined water can now be greatly enhanced. These devices have enabled safer operation and quicker turnaround times, hence improving the operational efficiency of ships.

Often, improvement gained on one aspect of the design would have a negative impact on another, for instance, in reducing motion with stabilisers, there are the penalties of extra machinery space and an increase in resistance. For the extra investment, the designer has to justify its use, and the likely benefits to be derived. Often, a designer needs to compromise over conflicting requirements of a design. In recent years, operational aspects are receiving greater attention from the designer. For instance, in bulk carrier safety, the cargo loading practices for this ship type may cause large stresses in the hull structure. Designers need to re-examine these loading practices in order to minimise their impact. Furthermore, as a system, the ship will eventually come to the end of its economic life. When it goes for scrapping, how much of it can be recycled and would any of the dismantled sub-systems cause harm to the environment? Naval architects will need to address these questions as new international practices and legislation demand.

Engineering economics is a major factor in ship design. It is not enough simply to reduce resistance or improve sea-keeping qualities. The ship must be viewed in its economic environment over its lifetime and optimisation must be related to many conflicting requirements. Ship design must be considered as the link between marine transport requirements and the most economic means of meeting them. In essence, the naval architect needs to produce a well balanced techno-economical solution to satisfy the shipowner's requirements.

The modern ship design process may proceed from an initial investigation of transport demand or the results of a market survey. A number of technical solutions might be offered, in terms of sizes and types of ships operating at different speeds. These designs might be reviewed in terms of first cost, operating cost and their respective incomes, ie, an economic evaluation. The designer should be familiar with, or aware of, a number of major cost aspects:

(a) Capital cost: design costs, building costs, loan interest/finance charges.
(b) Ship operating costs: range from fuel consumption, repair and maintenance, manning costs, cargo-handling, port charges and canal dues, to insurance and stores.

The ship designer would, in general, be more concerned with cost items in (a) and part of (b). For instance, the selection of certain machinery may impose a high capital cost but may reduce maintenance cost, while the use of stabilisers will increase capital cost, but would improve the operational effectiveness. Also, the designer should have an appreciation of his or her design's impact on production costs, and aim to reduce the production cycle time and material wastage.

12.2 Design category

Ships can be broadly categorised into three types for design purposes by considering either mass, volume or linear dimensions. Each category has different dominant features in their design. When the cargo mass is high, compared to its volume, the design is governed by the weight of cargo. For example, iron ore, which is a dense cargo, is usually transported by bulk carrier. Such ships, when fully loaded, would float close to their minimum freeboard. If the volume of the cargo is high compared to its mass, ie for lighter cargoes, the design is governed by space (capacity) requirements, eg containerships or passenger vessels. These ships, when fully loaded, would have a lot of freeboard. Where a ship's dimensions are limited by external constraints, such as water depth or canal width, then it is in the linear dimension category. Sometimes, external constraints could be imposed by terminal facilities, or by cargo size, eg containers. In practice, many ships can be a combination of all three categories, in which both weight and volume have to be considered, along with dimension constraints.

12.3 Design process

Ship design is a creative process, with due regard given to performance and economic objectives, through which a definition of the ship is developed, prior to its construction. The overall process is a number of stages, as depicted in Fig 12.1, which begins with the owner's requirements. Although shown as a sequential process, in practice there would be interaction between the different stages.

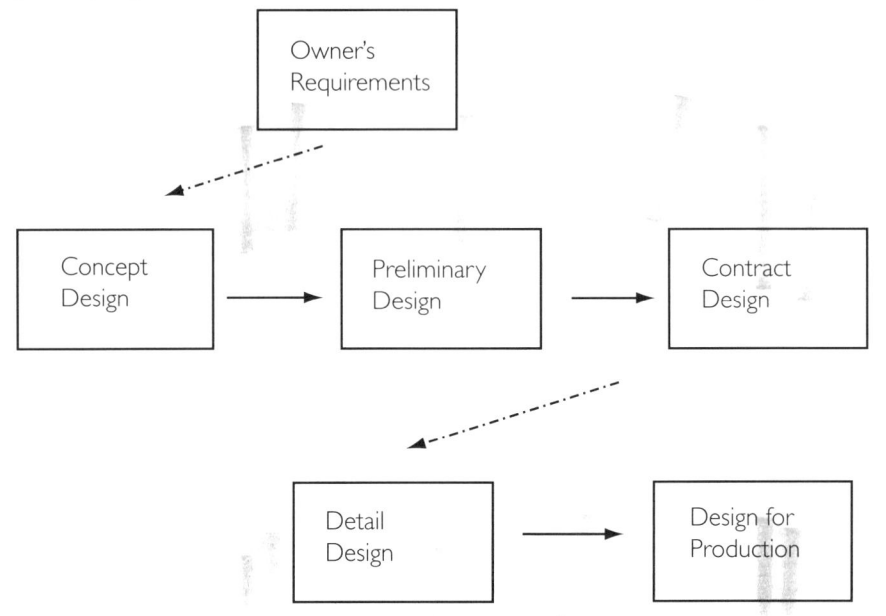

Fig 12.1: The ship design process

Concept design deals with the initial trial-and-error attempts to produce the principal characteristics of a ship that would meet the owner's requirements. The next phase is **preliminary design**, in which more detailed calculations will be carried out, refining the hull form to an optimum design. **Contract design** then deals with the development of contract plans and specifications, which provide a basis for tendering for construction offers from shipyards. Conceptually, the ship design process can be considered as a design spiral, as outlined by Buxton.[1] The concept is illustrated in Fig 12.2. A design begins with certain basic data and a cycle is undertaken which spirals out from the centre. As the cycle is repeated, fewer ship designs are considered, many having been rejected for reasons both technical and economic. More detail is developed for the smaller number of designs, as the spiral moves outwards. Thus, an optimum ship design is arrived at through successive iterations of design parameters.

At the **detail design** stage, it is the responsibility of the shipyard awarded the contract to develop the plans for building the vessel. Design for production will be a further elaboration of the work to prepare for construction. As the design progresses from one stage to another, the amount of time spent would escalate exponentially, from hundreds to ten of thousands of man-hours.

Merchant Ship Naval Architecture

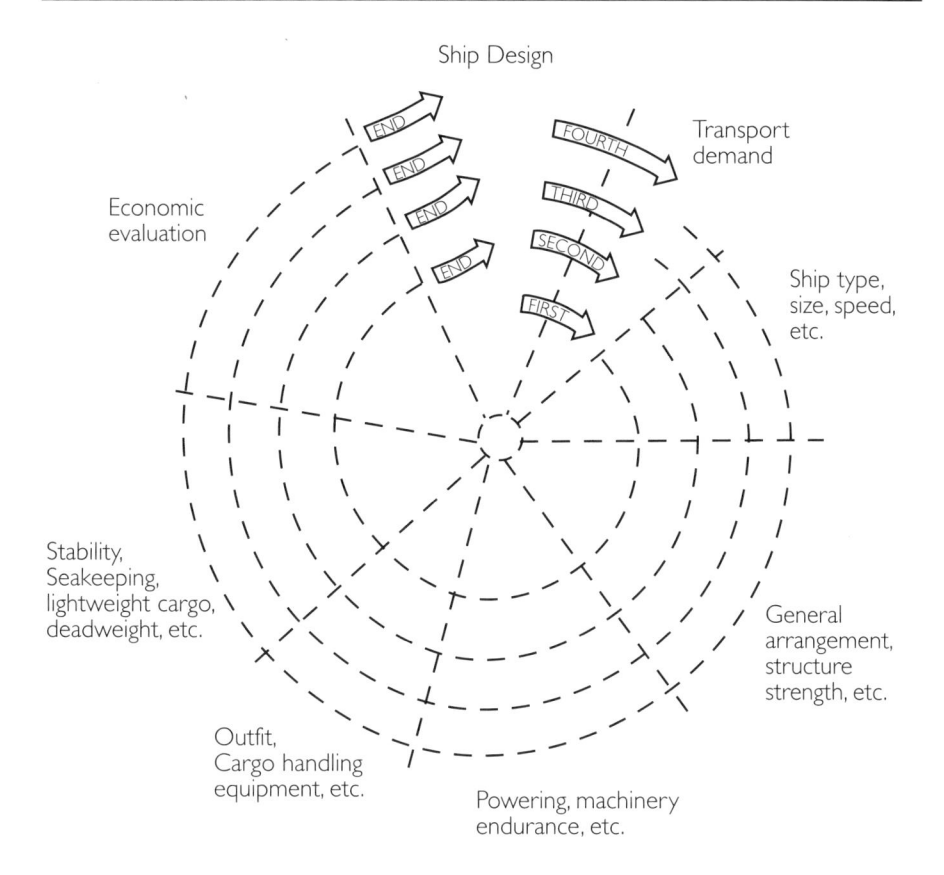

Fig 12.2: A ship design spiral

The role of the naval architect will change as the ship design process unfolds. In the conceptual design, they must communicate clearly to the shipowner what the best options are, within the constraints of technological and economic requirements. Close collaboration between the owner and designer will ensure the smooth running of this phase of the project. The second phase, preliminary design, is aimed at the engineering and design management aspects, in which the design team is to make sure that all aspects of the design are properly examined, to meet performance targets. At the contract design phase, the designer is likely to be in touch with the prospective shipbuilder, ensuring that the design work can be carried out effectively during the construction phase.

12.3.1 Concept design

This is the first phase of the overall ship design process, in which basic design requirements in terms of speed, range, cargo volume or weight, are translated into parameters that the naval architect can build up in their design framework. These parameters are the principal dimensions, eg length, beam, depth, draft, fullness and power, of one or a number of proposed ships. Conceptual designs are broad-brush in nature, with plenty of room for creativity and divergent thinking. This stage of the design should bring in ideas from new developments in design technology, changes in regulations, new materials and fabrication technology.

Through an iterative process, that allows variation in displacement, fuel consumption, stability and costs to be analysed, optimum designs can be obtained. This process can be depicted as the inner first loop of the design spiral, see Fig 12.2. In practice, constraints imposed by the prevailing economic situation could negate the viability of certain designs. In a more traditional approach, it is quite common to start with a design that has satisfied similar operational requirements. By revising the existing design, a more efficient and economical ship could be obtained.

For the initial sizing of a ship, if the required displacement is known, this can be related to the principal dimension, using the block coefficient:

$$\Delta = \rho C_B L B T$$

where ρ is density, L is length of the ship, B is breadth, T is the load draught; and C_B is the block coefficient. In theory, many combinations of length, breadth and draught can satisfy the displacement requirement of the basic design. In practice, the ship type and the speed expectation would limit the choice of C_B. For instance, a fuller block coefficient will be used for a deadweight carrier, eg a bulk carrier, which would be approaching 0.8 and may even exceed this value. For a containership, which is of a finer form, a C_B of about 0.6 would be common.

As the block coefficient increases, resistance and, hence, propulsion power, would increase and at a much faster rate. At some point, the power required would become uneconomical for driving such a full form ship.

For the three dimensions, at this design stage, the normal range of L/B and B/T ratios of the ship type under considerations can be used for initial estimations. In general, based on stability considerations, the B/T ratio would need to be above 2.25 and for lower resistance, the L/B ratio would need to be as high as practicable. Through these ratios, the displacement can be related to its length, from which an estimation of length, L, can be obtained. For example, for a modern bulk carrier of block coefficient 0.85 which is to have a displacement of 94 000 tonnes, where L/B is 7.5 and B/T is 2.4, the equation becomes:

$$\Delta = \rho C_B L(k_1)(k_2) \text{ where } k_1 = B/L, k_2 = (B/L)(T/B)$$

$$94\,000 = 1.025 \times 0.85 \times L^3 (1/7.5)(1/7.5)(1/2.4)$$

hence, L = 244.2m, B = 32.56m, T = 13.5m

With this draught, the ship would be able to transit the Panama Canal. However, the breadth would be too wide for some narrower waterways.

The next stage would be to estimate the freeboard, based on the load line regulations. This can only be done by a process of trial and error, searching for the depth that would give the desired draught. The length-to-depth ratio derived would then need to be checked, to ensure it meets the minimum standard with some margin. It must not be unduly large, since this may have an impact on other design parameters, eg structural considerations. For the bulk carrier above, the freeboard would be about 5.1m, while the minimum freeboard would be 5.03m.

The designer would also want to know about the mass of the ship, which is made up of the steel hull, the superstructure and deckhouses, equipment and outfit, and the main engine. This will normally be estimated, based on similar designs, or empirical formulae. For the bulk carrier above, the light-ship weight would be about 16 800t, of which steel weight makes up about 13 700t.

12.3.2 Preliminary design

Preliminary design can be considered as the second and further loops in the design spiral. At this stage of the design, more detailed analyses and calculations are required in the design process. Additional data would be needed to check results and the validity of earlier assumptions. Ship's lines, body plan and general arrangements would be developed for the final design.

From the ship lines, an optimum hull form, with acceptable performance, can be determined at this stage. This could be done by selecting a mother ship form from the methodical series, or data from an existing sister ship. Based on a roughly-faired preliminary body plan, basic hydrostatic data, sub-division and freeboard can be generated, using specialist computer programs.

More in-depth analyses will be carried out on resistance and powering, sea-keeping, and manoeuvrability, such that the motions in certain sea states and the turning circle of the vessel are ascertained. Structural profiles and arrangements would be done and an initial prediction of potential vibration problems would also be useful at this point. When requirements are broadly met, the ship's lines will be finalised, before model testing for predicting actual performance. A wake survey for propeller design will also be conducted.

General arrangement drawings will be further refined, after which final structural details will be determined for the midship section, the scantlings of structural parts to meet classification society rules for strength estimation. Bending moment and shear force distribution, structural efficiency, design of transverse bulkheads and structural continuity will have to be examined thoroughly. Often finite element analysis would be carried out, in particular if details of a particular aspect were required, eg the torsional response of a containership. Further vibration response analysis of the ship would also be carried out. The designer must minimise weights and consider their distribution in every design cycle.

For each of these design aspects, there would be an optimal performance. However, this is only a local optimal solution of an overall design; it would need further examination from other technical, or even economic perspectives, ensuring an overall optimum design is achieved. For instance, in structural optimisation, the criterion of optimisation may be that of a minimum weight. This would have a positive impact on cost, however, in an extreme situation, as in the case of bulkhead spacing, this might be inconsistent with minimum bulkhead spacing for damage stability purposes. It is important for the designer to have an appreciation of the sensitivity of the variables that have a direct influence on this criterion and other constraints that may be imposed. For further discussions on ship design optimisation, readers are referred to Schneekluth[2].

12.3.3 Contract design

Once the owner agrees to the preliminary design, the design will need to be produced to a level of detail from which the ship can be ordered, the contract design. All major features of the ship will be fixed, and drawings and specifications will be available for the development of the final design. This forms an integral part of a shipbuilding contract, including for instance, a fair set of lines, machinery arrangements, general arrangements, steel-scantling plan and performance specifications. The shipowner will then tender out to shipyards for offers to build. Detailed negotiations may take place with a small number of selected shipyards and then a contract to build will be signed. The shipyard will then take over the design process and create detail drawings to suit its manufacturing facilities.

12.4 Technology impacts

There are two main aspects to the impacts of technology on ship design; the use of computers in the design process, and developments in new marine technology. These have enabled the creation of new ship types, such as trimarans, or enhancing and expanding the performance of conventional ships, eg, the azimuthing propulsion unit for passenger ships and other vessels.

The greatest impact of computers in the design process has been the speed with which design and analysis can be carried out. This has allowed many variations in design parameters to be studied in a short space of time. With experience accumulated, the design cycle would be shortened. Furthermore, it would also allow greater innovation when compared to the traditional prescriptive rules and experimental approach using physical models. The power of this numerical approach is further enhanced when specialist areas are integrated to form a suite of programs, allowing data to be transferred from one to another area of application. In Fig 12.3, a typical example of an integrated suite of software for the ship designer is outlined.

Another important feature of computer application is the ability to visualise the spatial perspective of hull forms and internal arrangements. A virtual prototype of the ship can be created with great speed and ease. If ship design is described as an iterative process, the computer is offering the designer an interactive approach to design. The graphical interface renders great assistance to the designer and prospective shipowner alike. With advances in computer technology, both in terms of hardware and software, the idea of a 'numerical towing tank' may soon become a reality. In recent years, a knowledge-based approach to design, and the use of design database methodology are beginning to add further value to the design process. The link between design and production has also been greatly facilitated by computer.

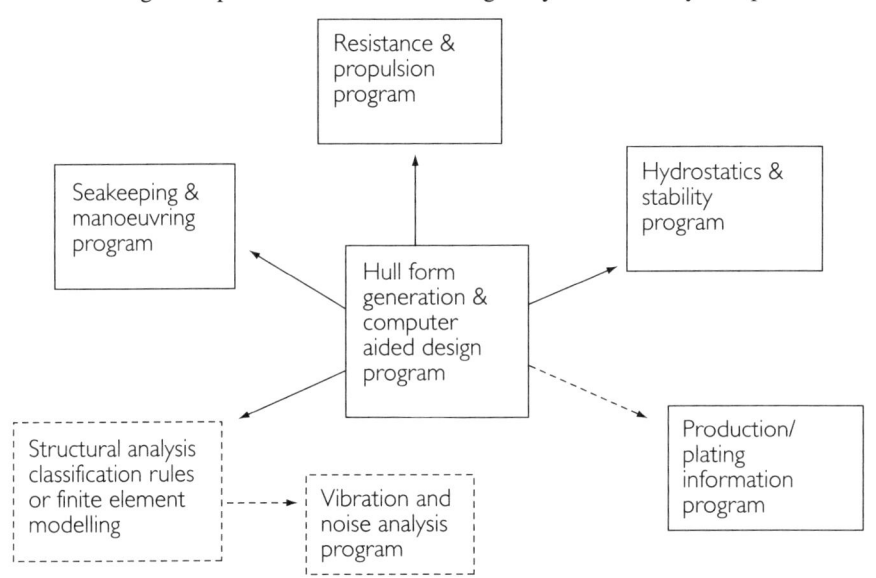

Fig 12.3: Typical integrated ship design software

In short, ship design is not a static process, but one that is evolving constantly to meet new frontiers created by technological developments and market demands. It has moved on from a traditional approach that places a heavy reliance on empiricism, to one that is based on sound, scientific, principles and engineering analysis. On this basis, it is envisaged that innovation will flourish, and a case in point is the rapid development of high-speed ferries over the past decade.

Whilst much of the emphasis has been placed on technical excellence and meeting performance specifications in ship design, it is essentially a techno-economic process. The ship design process may account for 5% of the ship's purchase cost, but it will influence over 70% of the ship's operating costs over its lifetime.

12.5 References

1. Buxton IL. *Engineering economics and ship design*, BSRA, 1976 edition.
2. Schneekluth H. *Ship design for efficiency and economy*, Butterworths & Co, 1987.

Index

A
Added mass (vibration) 184
Added weight (flooding calculation) 66, 67
Aframax 7
After perpendicular 17
Amidships 17
Angle of loll 52
Angle of repose 49
Archimedes' principle 39
Attwood's formula 50, 53
Augment of resistance (ship model) 151
Axial inflow factors (propeller) 138
Azipod propulsion unit 157

B
BACAT 6
Back (propeller) 133
Barge carriers 5
Base line 17
Beaufort Scale 73
Bending moments 86, 89
Bilge keels 170
Bilge radius 18
Block coefficient, C_B 19, 65
Blockage factor 65
Body plan 18
Bonjean curves 40, 70, 89
Boundary element (propeller) 141
Boundary layer 117, 120
Brittle fracture 115
Bulk carriers 7
Bulkhead deck 66
Buoyancy 39

C
Camber 18
Capesize vessels 9
Catamarans 6, 15
Cauchy number 118
Cavitation 152
Cavitation bucket diagram 153
Cavitation number 118, 152, 153
Cavitation tunnel 154
Centre of buoyancy, B 40
Centre of flotation, F 36

Centre of gravity, G 40, 43
Centre of pressure (rudder) 159
Chemical tankers 12
Chord (propeller) 135
Classification societies 23
CLT propeller 157
Coamings 2
Combination carriers 7
Compartmentation 71
Concept design (ship) 202, 203
Containers 3, 4
Containerships 3
Contract design (ship) 202, 205
Contra-rotating propellers 157
Controllable pitch propeller 134, 155
Corresponding speeds 120, 126
Criterion of service 71
Cruise liners 14
Curves of statical stability 50, 53
Cycloidal propellers 155

D

Deadweight 21
Deck Line 28
Detail design (ship) 202
Developed outline (propeller) 135
Displacement 21, 39
Displacement determination 61
Draught 18
Drydocking 62
Ducted propeller 156
Dynamical stability 56
Dynamic loading 87

E

Entrance 19
Environmental elements 73
Even keel 56
Extreme breadth 17
Extreme depth 17
Extreme draught 17

F

Face (propeller) 133
Factor of subdivision 71
Fairness 18
Fast Ferries 14

Ferries 14
Finite element analysis 114, 186
Fin stabilisers.. 171
Flare.. 18
Flat of keel 18
Flexural vibration.. 180
Floodable length 69
Floodable length curves 70
Flooding... 66, 69
Forward perpendicular... 17
Freeboard.. 18, 28, 82
Free surface 46
Fresh water allowance 62
Frictional resistance... 120
Froude number.. 118
Froud's Law of Comparison 120, 126
Fully cavitating propeller....................................... 153

G

General cargo ships 2
Grim wheel 157
Gross tonnage 21, 29

H

Half breadth plan 19
Handymax vessels 9
Handysize vessels 9
Hatch covers 2
Hatchways 2
Heave.. 78
Heeling forces (ship turning).................................... 164
Higher order vibration 191
Holds 2
Hogging 62, 86
Hovercraft 15
Hull efficiency 148
Hull efficiency elements 149
Hull form 18
Hydrofoil craft... 16, 173
Hydrostatic curves 59

I

Ideal efficiency (propeller) 138
Immersion 57
IMO Conventions 6, 7, 9, 12, 23, 26, 29
Inclining experiment 43, 44
International Maritime Organization (IMO) 24, 25

Isochronous rolling 79

K
Kumai's formula (vibration) 185

L
LASH 5
Length between perpendiculars, LPP 17
Lifting line (propeller)... 141
Lifting surface (propeller) 141
Lift-on lift-off 5
Lightweight 21
Liquefied gas carriers 9
Liquefied natural gas carriers 10
Liquefied petroleum gas carriers................................... 10
Lloyd's Register.. 23, 24
Load Line Convention.. 26, 28
Load line mark.. 28
Lo-lo 5
Longitudinal centre of flotation, LCF or F 36
Longitudinal metacentre, ML 57
Longitudinal metacentric height, GML............................. 57
Longitudinal stress 85
Lost buoyancy 66, 67

M
Mach number... 118
Margin line... 66
Maritime & Coastguard Agency (MCA), UK......................... 24
MARPOL 73/78 .... 6, 7, 9, 12, 27, 200
Metacentric diagram 43
Metacentric height, GM.. 41, 44
Midship area coefficient, CM 20
Midship section area, AM 20
Midship section diagram 92
Model tests... 108, 117
Moments of area 31, 34, 36
Moments of inertia 31, 92, 95
Moment to change trim, MCT...................................... 58
Motion control surfaces... 169
Moulded base line 17
Moulded breadth 17
Moulded draught 17

N
National authorities 24
Net tonnage 21, 29

Neutral point (ship) 159
Noise 195
Noise control 196
Noise prediction 196

O
OBO 8
Oil Pollution Convention 27
Oil tankers 6
One-compartment ship 70
Ore/bulk/oil carriers 8
Ore carriers 8
Ore/oil carriers 8
Overload fraction (propeller) 150

P
Panamax vessels 7, 9
Parallel middle body 19
Passenger ships 13
Period of encounter 75
Permeability 66
Permissible length 71
Pitch (propeller) 133
Pitch (ship) 78
Pounding 82, 99
Preliminary design (ship) 202, 205
Prismatic coefficient, CP 19
Products carriers 7
Projected outline (propeller) 135
Propeller coefficients 43
Propeller design procedure 151
Propeller efficiency 149
Propulsive coefficients 149

Q
Quasi-propulsive coefficient 149

R
Rake 8
Range of stability 0
Rayleigh distribution 6
Rayleigh's law 119
Refrigerated cargo ships 3
Relative rotative efficiency 48
Reserve buoyancy 66
Resistance .... 118 *et seq*
Response amplitude operator (waves) 81

Reynold's number 118
Righting lever, GZ.. 41, 46
Righting moment 41
Rise of floor 18
Roll-on roll-off (ro-ro)ships.. 4, 14
Roll (ship) 78
Rotational inflow factor (propeller)................................. 138
Rudders 159
Rudder types .... 161, 167 *et seq*
Run 19

S
Safety of Life at Sea (SOLAS) Convention 26, 27, 70, 71
Sagging 62, 86
Sails 158
Schlick formula (vibration) 185, 189
SEABEE 5
Seakeeping... 82
Seasickness 83
Seaworthiness 82
Shear force... 86, 89
Sheer 18
Sheer plan 18
Ship design process 202
Ship design software 206
Ship dimensions 17
Ship-model correlation 122, 127
Ship movement 78
Ship performance... 83
Ship vibration.. 179
Significant wave height 76
Simpson's rule 31
Singing (propeller)... 152
Slamming.. 82
Small Waterplane Area Twin Hull (SWATH) 16
Smith effect 75
Sound intensity 195
Sound measurement... 195
Spectral density (waves) 77
Speed trials... 154
Spiral manoeuvre... 165
Springing (vibration)... 179
Squat 64
Stabilisation 83
Stability 1, 46, 48, 49, 56, 63, 69, 83
Standard series data (hull form) 129
Static loading 85

Stations 18
'Stiff' ship 41, 79
Strength .... 82, 85 *et seq*
Stresses .... 85 *et seq*
Subdivision bulkheads................................... 66
Subdivision criteria 70
Subdivision index 71
Suezmax vessels 7, 9
Super cavitating propeller 153
Surface effect ships 15
Surge 78
Sway 78

T
Tank divisions 47
Tank-type stabilisers 172
Technology impacts (ship design)...................... 206
'Tender' ship 41, 79
TEU 4
Thrusters 157
Todd's approach (vibration)........................... 189
Tonnage 21, 29
Tonnage systems 29
Tonnage Measurement Convention 26, 29
Tonnes per centimetre immersion, TPC 57
Torque (rudder stock) 161
Torsional vibration................................... 180
Towing tanks 121
Transverse stability 40
Trapezoidal rule 31
Trim 18, 56
Trimming lever 60
Trimming moment....................................... 58
Tumblehome 18
Turning circle (ship) 163, 165
Two-compartment ship 70

V
VLCC 6
Vertical centre of buoyancy, KB 40
Vertical prismatic coefficient, CPV 20
Vibration calculation methods (direct) .... 192 *et seq*
Vibration damping 176
Vibration formula (ship) 183
VLCC 6

W

Wall-sided formula 51
Wall-sided ships 50
Water density changes 62
Water density values 73
Waterjet propulsion 158
Waterplane area, AW 20
Waterplane area coefficient, CW 20
Wavemaking resistance 125
Waves 73, 80
Weber number 118
Wetted surface area 124
Whipping (vibration) 179

Y

Yaw (ship) 78

Z

Zig-zag manoeuvre 166

Tonnage measurement section 3.9
AMENDMENT
to equation at the top of page 30
$K_2 = 0.02 + 0.02 \log_{10} V_c$
should read:
$K_2 = 0.2 + 0.02 \log_{10} V_c$